W9-ADO-869

Praise for *Brain Energy*

"*Brain Energy* is a dramatic breakthrough in understanding mental illness by a leading Harvard psychiatrist who provides a revolutionary road map for people suffering from depression, anxiety, bipolar disease, in fact, almost any brain disorder. The discoveries of brain science have forced us to reimagine mental health and provide hope in a field that for many has not provided significant relief from suffering. If you suffer any mental health challenges (who hasn't), then this book just might change your life."

—Mark Hyman, MD, senior advisor at the Cleveland Clinic Center for Functional Medicine and 14-time *New York Times* bestselling author

"*Brain Energy* provides a long-awaited unifying mechanism underlying a vast spectrum of mental illness conditions. And this new paradigm will undoubtedly usher in potent therapeutic interventions for pervasive psychiatric conditions for which standard pharmaceutical approaches have proven minimally effective. Christopher Palmer's work empowers patients and their health care providers alike."

—David Perlmutter, MD, #1 *New York Times* bestselling author of *Grain Brain*

"Dr. Palmer takes a provocative and insightful look into the origins of mental disorders, which have profound implications for how we treat the disease . . . and for our diet."

—Jason Fung, MD, nephrologist and author of three bestselling health books

"Psychiatry will never be the same. The medical profession needs to apply Biochem 101 if it's going to solve mental illness—and metabolic illness, while they're at it. Christopher Palmer poses the hypothesis, supports it with data, and, in my humble opinion, nails the treatment: feed the brain what it needs."

—Robert H. Lustig, MD, MSL, emeritus professor of pediatrics at UCSF and author of *Metabolical*

"For more than two decades, Dr. Palmer has organized world leaders in psychiatry to educate clinicians about emerging innovations in the field. He challenges the status quo with this brave new idea in *Brain Energy* that instructs us to seek simple solutions to human problems, rather than chemical ones that can only change biology, not lived experience. *Brain Energy* is a book all psychiatric professionals should read as a useful criticism of our field's major deficits. It is a book all people should read to understand how much they can do (and not do) for the sake of their mental health. Basic health is self-evident but takes courage to commit to taking exquisite care of your body and, by extension, your brain."

—Lois W. Choi-Kain, MD, MEd, director of the Gunderson Personality Disorders Institute and assistant professor of psychiatry at Harvard Medical School

"After a bipolar episode at age nineteen, our son was seen by more than forty mental health practitioners and prescribed twenty-nine different medications. But it was not until he started on a ketogenic metabolic therapy under the guidance of Dr. Chris Palmer that he got his mind and his life back. Dr. Palmer's metabolic approach has the potential to radically impact the world's mental health epidemic. *Brain Energy* is a must-read."

—David Baszucki, founder and CEO of Roblox and cofounder of the Baszucki Group, and Jan Ellison Baszucki, author of *A Small Indiscretion* and cofounder of the Baszucki Group

"If you have ever been dissatisfied with the rather hard-to-defend explanations of mental illness, this groundbreaking book is for you. Palmer, a practicing psychiatrist, not constrained by the received wisdom of the field, audaciously travels further than most. He puts forward a strong case for throwing away what we were taught about the causes, the diagnosis, and the treatment of psychiatric disorders. Instead, he brings the tiny mitochondria, once mere bacteria, into center stage and takes you on an exhilarating intellectual journey to reveal the new beginnings of psychiatry."

—Zoltán Sarnyai, MD, PhD, professor and head of the Laboratory of Psychiatric Neuroscience, James Cook University, Australia

"Dr. Christopher Palmer has written a must-read primer for anyone considering understanding and treating mental health. The book will guide you to understand why metabolism and mitochondria are fundamental to keep your brain healthy . . . a call to action to transform mental health treatment. Read this book—and learn from one of the best."

—Ana C. Andreazza, PhD, professor of pharmacology and psychiatry, University of Toronto and founder and scientific director, Mitochondrial Innovation Initiative

"Dr. Palmer is uncanny in his synthesizing extant literature and providing a prescient thesis on the pathoetiologic and potentially therapeutic role of metabolics of neuropsychiatric conditions. The thesis and framework proffered by Dr. Palmer provides for many prevention and therapeutic opportunities in psychiatry and take us one step closer to disease-modifying possibilities."

—Roger S. McIntyre, MD, FRCPC, professor of psychiatry and pharmacology at the University of Toronto, Canada

"*Brain Energy*, by Dr. Chris Palmer, is *the* much-needed new perspective on mental health that could revolutionize the way we think about, research, and treat mental health conditions. It's a guaranteed best-seller, a book you won't want to put down, and one that could change your life or that of a loved one. The nuance, clarity, and sensitivity with which Dr. Palmer addresses the complex issues of mental illness is nothing short of incredible. He brings his skills as a Harvard-trained clinical psychiatrist to bear in his writing, anticipating the reader's next question and then weaving it into a narrative that is the perfect mix between science lesson and storybook. Filled with brilliant analogies, jaw-dropping statistics, fascinating scientific details, and moving patient stories, this book is an absolute must-read. And, speaking as both a PhD scientist myself and a Harvard medical student, I wish this book could be mandatory reading for the next generation of doctors. If it were, I'd suspect we'd see rates of metabolic diseases, including mental health conditions, start to dip within a generation. This book will change many lives."

—Nicholas Norwitz, PhD in neurometabolism (University of Oxford) and Harvard Medical Student

"There is so much we do not know about the relationship among metabolism, health, and disease. Thankfully, Dr. Christopher Palmer cuts through the noise and tackles mental disorders by focusing on first principles: metabolism is the foundation of brain health, and brain health is the foundation of our best future. *Brain Energy* is a book that cannot be read and incorporated into policy soon enough: mental disorders are an accelerating crisis, especially among children. This is not rocket science; it's neuroscience."

—Susan A. Masino, PhD, professor of applied science and neuroscientist at Trinity College

"Kudos to Dr. Chris Palmer for penning a thought-provoking and superb book on the revolutionary breakthroughs occurring in psychiatry, a discipline in medicine that has suffered far too long and from too much stigma. It's a must-read for all of us as we undoubtedly, without exception, have a family member or a dear friend battling a challenging psychiatric condition. There is finally optimism and light at the end of the dark tunnel."

—Sanjiv Chopra, MBBS, MACP, FRCP, professor of medicine at Harvard Medical School and bestselling author

"It is clear to me from clinical practice that what we eat and drink affects our brain function and mental health but I never understood why. Dr. Palmer brilliantly connects the dots to explain why this is true. A pioneering work!"

—Eric C. Westman, MD MHS, director of the Duke Keto Medicine Clinic

"Dr. Palmer's *Brain Energy* seemed to take me by the hand and gently walk me through the most complex of medical journeys—arriving at the most fascinating yet incredibly logical conclusions. Though not a scientist, I was able to follow and actually become riveted during every step of the journey. His rare ability to explain sophisticated scientific and medical concepts in lay terminology makes it accessible to a very wide audience. His case is carefully and thoroughly constructed. The dots are always connected. His frequent use of anecdotes and examples is also helpful. Visualizing metabolism by comparing it to automobile traffic is kind of brilliant. Turning the *Titanic* of the understanding of mental health from genetic to metabolic underpinnings is not going to be easy, but *Brain Energy* seems to be a huge swing in the right direction."

—Jim Abrahams, director of The Charlie Foundation for Ketogenic Therapies

"Not since the foundational theories of psychodynamics (Sigmund Freud) and behaviorism (John Watson) has a bold and potentially transformative new proposition emerged to explain the mounting epidemic of mental illness throughout the world and across the age-span. Dr. Chris Palmer's groundbreaking theory is that abnormalities in brain energy metabolism are likely root causes of psychiatric conditions and that dietary and metabolic approaches can be exploited to benefit patients, and even the general population at large. *Brain Energy* is a must-read for everyone interested in brain and mental health."

—Jong M. Rho, MD, professor of neurosciences and pediatrics at the University of California San Diego

Brain Energy

A Revolutionary Breakthrough in Understanding Mental Health—and Improving Treatment for Anxiety, Depression, OCD, PTSD, and More

Christopher M. Palmer, MD

BenBella Books, Inc.
Dallas, TX

This book is for informational purposes only. It is not intended to serve as a substitute for professional medical advice. The author and publisher specifically disclaim any and all liability arising directly or indirectly from the use of any information contained in this book. A healthcare professional should be consulted regarding your specific medical situation.

BenBella Books, Inc.
10440 N. Central Expressway
Suite 800
Dallas, TX 75231
benbellabooks.com
Send feedback to feedback@benbellabooks.com

BenBella is a federally registered trademark.

Printed in the United States of America
10 9 8 7 6 5

Library of Congress Control Number: 2022019354
ISBN 9781637741580 (hardcover)
ISBN 9781637741597 (electronic)

Editing by Alexa Stevenson and Greg Brown
Copyediting by Scott Calamar
Proofreading by Lisa Story and Ariel Fagiola
Indexing by WordCo Indexing Services, Inc.
Text design and composition by Aaron Edmiston
Cover design by Faceout Studio, Jeff Miller
Cover image © Shutterstock / ganjalex

To my mother

My futile attempts to save you from the ravages of mental illness lit a fire in me that burns to this day. I'm sorry I didn't figure this out in time to help you. May you rest in peace.

Contents

PART III: CAUSES AND SOLUTIONS

Introduction

For more than twenty-five years as a psychiatrist and neuroscience researcher, I have been asked the question "What causes mental illness?" countless times by patients and their family members. When I first began my career, I would give long answers that made me sound educated and competent. I'd talk about neurotransmitters, hormones, genetics, and stress. I would describe the treatments we would be using and offer hope that they would make things better. After a few years of this, however, I began to feel like a fraud. You see, people often weren't getting that much better. Treatments would sometimes work for a few months, or even a year or two, but more often than not, the symptoms would come back. At some point, I began telling people the simple truth: "No one knows what causes mental illness." Although we understand many risk factors, no one knows how they all fit together. I still tried to offer hope by assuring people that we had many different treatments at our disposal and that we would try one after another until we found one that worked. Sadly, for many of my patients, we never did.

That all changed for me in 2016 when I helped a patient lose weight. Tom was a thirty-three-year-old man with schizoaffective disorder, a cross between schizophrenia and bipolar disorder. He had suffered from hallucinations, delusions, and mental anguish every day of his life over the last thirteen years. He was tormented by his illness. He had tried seventeen different medications, but none had worked. The medications sedated him, which reduced his anxiety and agitation, but they didn't stop his hallucinations or delusions. What's more, they'd caused him to gain over one hundred pounds. He had long been plagued by low self-esteem, and being so overweight only added to this. He had become a near hermit, and our weekly sessions were some of his only excursions into the outside world. This is partly why I agreed to help him lose weight: I was the doctor he saw most often, and he wasn't in the market for a referral to a specialist he'd never met. More to the point, it was highly unusual for him to take action to improve his health in some way. Maybe losing weight could help him gain a sense of control over his life. After experimenting with several approaches without success, we decided to try the ketogenic diet—a diet low in carbohydrates, moderate in protein, and high in fat.

Within weeks, not only had Tom lost weight, but I began to notice remarkable and dramatic changes in his psychiatric symptoms. He was less depressed and less sedated. He began making more eye contact, and when he did, I saw a presence and spark there that I had never seen before. Most astonishingly, after two months, he told me that his longstanding hallucinations were receding and that he was rethinking his many paranoid conspiracy theories. He began to realize that they weren't true and probably never had been. Tom went on to lose 150 pounds, move out of his father's home, and complete a certificate program. He was even able to perform improv in front of a live audience, something that would have been impossible for him prior to the diet.

I was flabbergasted. I had never seen anything like this in my entire career. While it's possible that losing weight might reduce anxiety or depression in some people, this man had a psychotic disorder that had resisted more than a decade of treatment. Nothing in my knowledge or experience suggested that the ketogenic diet would treat his symptoms. There seemed to be no reason it should.

I began digging into the medical literature and discovered that the ketogenic diet is a longstanding, evidence-based treatment for epilepsy. It can stop seizures even when medications fail to. I quickly realized an important connection—we use epilepsy treatments in psychiatry all the time. They include medications like Depakote, Neurontin, Lamictal, Topamax, Valium, Klonopin, and Xanax. If this diet also stops seizures, maybe that's why it was helping Tom. Based on this additional information, I began using the ketogenic diet as a treatment with other patients and when it continued to be successful, I soon found myself collaborating with researchers around the world to explore it further, speaking globally on this topic, and publishing papers in academic journals demonstrating its effectiveness.

I set out on a journey to understand how and why this diet worked for my patients. Along with its use in epilepsy, the ketogenic diet is also used in treating obesity and diabetes, and is even being pursued as a treatment for Alzheimer's disease. At first, this was confusing and a bit overwhelming. Why would one treatment work for all of these disorders, even if only in some people? Ultimately, it was this question that opened the door to something much bigger than the inquiry I'd begun with. It forced me to uncover the connections between these different disorders and integrate this understanding with everything that I already knew as a neuroscientist and psychiatrist. When I finally put all the pieces together, I realized that I had stumbled upon something beyond my wildest dreams. I had developed a unifying theory for the cause of all mental illnesses. I call it the theory of brain energy.

This is not a book about the ketogenic diet—or any diet at all. It's also not solely concerned with serious mental illness; the scientific insights in this book apply to mild depression and anxiety as well. In fact, it may change the way you think about all human emotions and experiences. I am not offering a simple cure-all for mental illness, or advocating for any single treatment. The unexpected effectiveness of that particular treatment was only the first clue that started me on the path to a new way of understanding mental illness. This book will share that understanding with you, taking you on a journey that I hope will transform the way you think about mental illness and mental health.

Here's a quick overview of what's in store:

- I'll begin by reviewing where we are now in the mental health field: the problems and questions that plague us, and why they matter.
- You'll learn something that may seem shocking—mental disorders are not distinct entities. This includes diagnoses like depression, anxiety, PTSD, OCD, ADHD, alcoholism, opioid addiction, eating disorders, autism, bipolar disorder, and schizophrenia. There is tremendous overlap in symptoms for different disorders, and many people are diagnosed with more than one. And even among disorders with symptoms that are very different, the underlying biological, psychological, and social factors overlap significantly.
- I'll explore the surprising connections between mental disorders and some physical ones such as obesity, diabetes, heart attacks, strokes, pain disorders, Alzheimer's disease, and epilepsy. In order to truly understand what causes mental illness, these connections need to be understood as well.
- This will all come together to reveal that *mental disorders are metabolic disorders of the brain.*
- In order to understand what this means, you'll need to understand metabolism. It's a lot more complicated than most people realize, but I'll do my best to make it as simple as possible. Tiny things called mitochondria are key. Metabolism and mitochondria can explain all the symptoms of mental illness.
- I'll discuss the differences between normal mental states and mental disorders. For example, we all experience anxiety, depression, and fear at different times in our lives. These experiences aren't disorders—they are a normal part of being human. However, when these things happen at the wrong time or in an exaggerated way, they may cross the line from *mental state* to *mental disorder*. You'll

see that all mental states, even normal ones, relate to metabolism. For example, "stress" is a mental state that affects metabolism—it takes a metabolic toll. If it occurs for prolonged periods of time or is extreme, it can lead to mental illness. But so can anything else that affects your metabolism.

- I will share with you five broad mechanisms of action that can explain the clinical and neuroscientific observations we see in all mental disorders.
- I'll show you that all the known contributing factors to mental illness, including things like genetics, inflammation, neurotransmitters, hormones, sleep, alcohol and drugs, love, heartbreak, meaning and purpose in life, trauma, and loneliness, can be tied directly to effects on metabolism and mitochondria. I'll demonstrate how all these contributing factors affect metabolism, which then affects the function of cells, which can then result in symptoms of mental illness.
- You'll learn that all current mental health treatments, including the psychological and social ones, likely work by affecting metabolism.
- This new understanding of mental illness leads to new treatments, ones that offer the hope of long-term healing as opposed to just symptom reduction. They will sometimes be more difficult than just taking a pill, but they are well worth the effort. While more research will lead to additional new treatments, the exciting news is that many therapeutic options are available today.

To be clear, I am not the first to suggest that metabolism and mitochondria are related to mental illness. In fact, I am building on decades of research. Without these other researchers and their pioneering work, this book wouldn't exist. I'll share many of their groundbreaking studies in the pages to come. However, for the first time, this book puts the pieces of the puzzle together to reveal one coherent theory. This theory integrates existing biological, psychological, and social research, and offers one unifying framework for explaining and treating mental illness.

Brain Energy not only provides long-elusive answers, it offers new solutions. I hope it will end the suffering and change the lives of millions of people throughout the world. If you or someone you love is affected by a mental illness, it might just change your life, too.

Part I

Connecting the Dots

Chapter 1

What We're Doing Isn't Working

MENTAL HEALTH TODAY

The World Health Organization estimated that in 2017, almost 800 million people on our planet suffered from mental health disorders. This represents a bit over 10 percent of the world's population, or one in every ten people. When substance use disorders are included in the count, the number climbs to 970 million people, or 13 percent of the global population. Anxiety disorders were the most common, affecting about 3.8 percent of people around the world, followed by depression, affecting about 3.4 percent.[1] Rates of these disorders are higher in the United States, with approximately 20 percent, or one in five people, diagnosed with a mental or substance use disorder.

These numbers give us a snapshot of the prevalence of mental disorders over a specific one-year period. But lifetime prevalence rates are much higher. In the United States, data now suggest that about 50 percent of the population will meet the criteria for a mental disorder at some point in their lives.[2] Yes—*half of all people.*

Estimating the rates of mental illness is difficult. People often deny their mental health problems to others or even to themselves. Having a mental illness is stigmatized pretty much everywhere in the world. While societies have made important strides in recognizing things like depression and anxiety disorders as "real" illnesses, this progress is relatively recent, and it is far from universal. There are still people who see those suffering from these disorders as simply "whiny" or "lazy." On the other hand, while people with psychotic disorders are usually believed to have "real" illnesses, they face a different kind of stigma. Many people are afraid of them or dismiss them as "crazy." Then there are those with substance use disorders—not only do many see them as self-centered or morally weak, in some countries, such as some Middle Eastern ones, they are classified as criminals and can be incarcerated for using even alcohol. The effects of stigma can range from shame to outright discrimination, but stigma of any kind can motivate people to minimize or lie about symptoms. As such, prevalence statistics are most likely underestimates of the true scope of these disorders.

And as dire as these statistics are, the problem seems to be getting worse.

A Growing Epidemic

We have the best data for this in the United States, where researchers have been tracking mental health statistics for decades now. Rates of mental illness are on the rise. According to the CDC (Centers for Disease Control), US adults over the age of eighteen had higher rates of mental illness in 2017 than in all but three of the years between 2008 and 2015. Of note, the youngest group (those aged eighteen through twenty-five) had the largest increase—rising 40 percent between 2008 and 2017.

The rate of ADHD (attention deficit/hyperactivity disorder) is rising in children and adolescents, increasing 41 percent in children four through seventeen between 2003 and 2012. This particular diagnosis and its reported upward trend attract considerable controversy. Some suggest that we are simply getting better at recognizing this disorder and providing treatment to children who need it to thrive. Others suggest that we are medicating normal behavior—that society and schools have come to expect too much from children, and that our expectations are unrealistic for what they are capable of at certain ages. Still others argue that the attention span of the American population has decreased across the board, likely due to increased time spent in front of screens, and this is being mistaken for ADHD. Is the rate of this disorder truly increasing, or are these other factors responsible for what we see in the data? We'll further consider questions like this shortly. But ADHD isn't the only diagnosis on the rise.

Depression in children, adolescents, and young adults is increasing as well. From 2006 to 2017, rates of depression in the US increased by 68 percent in children ages twelve to seventeen. In people ages eighteen to twenty-five, there was an increase of 49 percent. For adults over the age of twenty-five, the rate of depression supposedly stayed stable.

However, much of this information is gleaned from surveys, and both the questions we ask and the way we ask them matters. Although surveys suggest that the rates of depression in adults are not increasing, many reports suggest that *burnout* is on the rise. Burnout is not an official psychiatric diagnosis in DSM-5 (the *Diagnostic and Statistical Manual of Mental Disorders*, Fifth Edition), but the World Health Organization recently added it to its list of mental disorders—the ICD-11 (International Classification of Diseases, Eleventh Revision). The criteria are similar to those for depression, but focus primarily on the stress of work and the work environment. There has been a lot of debate about whether burnout is simply a work-related form of depression, and for good reason: In one study looking at physician burnout, they found that those with mild burnout were three times more likely to meet criteria for major depression. Those with severe burnout were forty-six times more likely,[3] suggesting little, if any, difference between these diagnostic labels. Like depression, burnout is also associated with much higher rates of suicide. Because burnout is not yet an official diagnosis in the DSM-5, US agencies don't track its prevalence. However, a 2018 Gallup poll found that 23 percent of employees reported feeling burned out at work often or always, while another 44 percent felt burned out sometimes.[4] These rates are much higher than those of depression.

Suicide rates are increasing across most age groups. In 2016, nearly 45,000 people died by suicide in the United States alone. In general, for every person who dies by suicide, approximately thirty others attempt to kill themselves—putting the rate of suicide attempts at well over one million people per year. From 1999 to 2016, suicide rates increased in most US states, with twenty-five states seeing an increase of 30 percent or more. Another statistic, *deaths of despair,* tracks combined deaths in the United States from alcohol, drugs, and suicide. This statistic more than doubled between 1999 and 2017.

Anxiety disorders are the most common mental disorders, but the criteria for diagnosing them continue to evolve. This makes it difficult to assess changes over time. Some have argued that the rates have not changed in recent years.[5] However, an annual household survey of approximately forty thousand US adults suggests that anxiety is increasing. Survey participants were asked, "How often did you feel nervous during the past 30 days?" with five response options ranging from "all of the time" to "none of the time." From 2008 to 2018, rates of anxiety increased by 30 percent. In the youngest group, ages eighteen to twenty-five, there was an 84 percent increase.[6]

Sometimes the more "common" diagnoses like depression and anxiety are thought of separately from mental disorders like schizophrenia—the term "serious mental illness" is often used by mental health professionals to talk about disorders that involve significant impairment and disability, such as those with psychotic symptoms. While this category includes some severe forms of depression and anxiety, it mostly refers to diagnoses like schizophrenia, bipolar disorder, autism, and the like. So what about these disorders? What's happening with them? They're increasing, too. Between 2008 and 2017, there was a 21 percent increase in serious mental disorders in those over the age of eighteen in the United States. For the younger group, those aged eighteen to twenty-five, the rate of serious mental illness *doubled* during that same period—less than a decade.[7]

The diagnosis of autism is increasing at an alarming rate.[8] In 2000, autism affected about 1 in 150 children in the US; by 2014, it was about 1 in 59.

The statistics for bipolar disorder are also concerning. From the mid-1970s to 2000, the prevalence of bipolar disorder was somewhere in the range of 0.4 to 1.6 percent. By the early 2000s, it had increased to 4 to 7 percent.[9] In children and adolescents, this diagnosis was almost nonexistent prior to 1994, but it's increasingly common now.

These statistics are difficult to comprehend. Diagnoses like autism and bipolar disorder aren't supposed to increase exponentially over such a short period of time. While anxiety and depression can be situational, these other disorders are generally regarded as firmly "biological," and many researchers believe they are determined in large part by genetics. Clearly, the human species didn't have an epidemic of genetic mutations.

Researchers, clinicians, and society at large are struggling to understand what to make of the sharp increase in mental illness. While there is no consensus explanation, plenty of people have theories—and generally speaking, these theories can be divided into two categories.

The first category relies on the belief that the statistics are wrong, or that they don't mean what we think they do. Many people think it's impossible that the rates of mental disorders could increase so rapidly; they believe that these statistics are the result of doctors and/or patients seeing "disorders" that are not there. Here are three of the most prominent theories in this category:

1. It's the pharmaceutical companies! They are looking to sell pills to as many people as possible, and to sell pills, they have to convince both doctors and the public that they need them. They spend billions of dollars every year on marketing, sending samples to doctors to keep the name of their product top of mind. They run television ads that ask if the viewer has any of a variety of vague symptoms, like "decreased enjoyment." If you do, you are instructed to "talk to your doctor and see if 'drug X' is right for you." These types of

advertisements feed into people's tendencies to be hypochondriacs. These worried people then go to doctors and emerge with a new diagnosis, and of course, the pills to treat it.

2. It's laziness! People today don't want to work at things. They also don't want to experience any discomfort or think that they should have to. They increasingly categorize run-of-the-mill human emotions or experiences as "symptoms." They flock to therapists to relieve these "symptoms." Sometimes, they even go to the doctor to complain about them. People want quick and easy fixes; doctors are overworked and busy, and the easiest thing for them to do is write a prescription.
3. It's this new generation of kids! Given that the rates are increasing most dramatically in children and young adults, it is clear that the blame lies with them—or their parents. Parents have pampered and spoiled this younger generation—catering to their every whim. These children and young adults have never been disciplined and also don't have much willpower or perseverance. They get easily frustrated and overwhelmed. When their parents are no longer there to fix things, or when anyone tells them "no," they go into a crisis. These meltdowns get them diagnosed with some mental disorder. Or, unable to handle life in the real world, they look for a "mental illness" to blame.

As appealing as these types of theories may be, they are likely not the answers. If you do not suffer from a mental health issue yourself, or have a child who does, or have daily contact with patients who do, it's easy to assume that these people are simply whiners and complainers, and that doctors, patients, and parents are looking for quick fixes. It's easy to dismiss a problem when that problem is far removed from you. However, when you are confronted with the real people behind these statistics and witness their suffering, these blanket assumptions become impossible to sustain. When someone you know to be a "good parent" has a seven-year-old who suffers from fits of rage, who is not sleeping, who is threatening to kill himself or other people, the problem begins to look like a real one. These behaviors are not normal. When a woman has such severe panic attacks that she stops leaving the house, this is not normal. When a person is so depressed that they sometimes cannot get out of bed in the morning, this is not normal.

So, the second category of theories about the increasing rates of mental disorders accepts the statistics as real. These are people who believe that these illnesses are truly on the rise. They offer a variety of perspectives and possible explanations:

1. It's a good thing! These statistics are positive—they reflect a broader understanding of mental disorders and more awareness of how to identify them. There are many programs in schools and work environments on recognizing the symptoms of mental and substance use disorders. There are public service campaigns focused on suicide prevention. Celebrities are speaking out about their own mental health struggles, and there is more coverage of mental health in the media that is explicitly aimed at raising awareness *and* reducing stigma. People are increasingly getting the help they need, being diagnosed, and treated.

2. It's society! We have become increasingly dependent on technology and screens. As we sit and look at our phones, computers, or televisions, we have become both more sedentary and more isolated. We interact with each other less in "real life," connecting through social media instead of spending time together or talking on the phone. People only post the parts of their lives that "look good," so social media fosters unrealistic expectations and shame, not real connection. The pace of life is faster, too. Everyone is busy and overscheduled—even children. Families aren't having dinner together like they did in "the old days." No wonder people feel burned out. No wonder so many people are developing mental disorders.

3. It's toxins, chemicals, and fake food! It is not just the behavior of society that has changed, it is the physical world we live in. We are exposing ourselves to toxins every day. The foods we eat are filled with artificial ingredients. New chemicals are everywhere—on our lawns, in our water supply, in personal hygiene products that we use morning and night. We create and surround ourselves with compounds we'd never encounter in nature, the effects of which—especially in combination with so many other compounds—we do not fully understand. These are leading to increases in all sorts of illnesses, including cancer, obesity, and mental disorders, too, even if we don't yet know precisely how.

There are many more of these "second category" theories for the rise of mental health problems, but these are some of the most commonly discussed ones. None of these three are far-fetched. They might well play a role with at least some people, or at least some of the time. As I will explain later in this book, some of them almost certainly do.

But as for number one on the list above—rationalizing the statistics as simply the result of improved recognition and diagnosis—the evidence shows that it isn't just recognition that is increasing. The surveys comparing year-after-year data include samples of the entire population, regardless of whether people get diagnosed or not. These disorders are truly on the rise.

Perhaps the most important point to note is that the rates of very different mental disorders—autism, bipolar disorder, depression, and ADHD, to name a few—are *all* increasing at the same time. Why would that be? We think of bipolar disorder, ADHD, and depression as very distinct from one another, with distinct contributing causes. If these disorders are genetic, what happened to our genes? Is there a toxin causing numerous mutations? If the culprit is the stress of our fast-paced modern society, why are *all* of the disorders increasing? Wouldn't more stress simply lead to more depression and anxiety? Certainly, stress doesn't cause autism and bipolar disorder. Or does it? These statistics raise more questions than they answer.

To add insult to injury, the COVID-19 pandemic has taken an additional toll. In June 2020, an estimated 40 percent of all US adults reported struggling with mental health or substance use issues. Eleven percent of the adults surveyed said they had considered suicide in the past thirty days.[10]

The Price We Pay

Mental disorders are costly to society. The financial toll worldwide was $2.5 trillion in 2010 and is expected to reach $6 trillion by 2030.[11] These figures include the costs of direct mental healthcare services (hospitalizations, doctor and therapist visits) and prescription medications. But there are also other financial costs that are more difficult to measure, including lost productivity due to employees becoming less focused or taking sick leave. These losses affect employers and employees, societies and individual sufferers. Depression is now the diagnosis that tops the list of disabling illnesses—above all other illnesses, including cardiovascular disease, cancer, and infections. Mental and substance use disorders are the leading cause of "years lost to disability" and "overall disease burden" in the US.[12]

Much more important than the financial costs of mental disorders is the suffering they cause to individuals and their families. They cause untold misery and despair. Mental illnesses can ruin people's lives. They may lead to social isolation, disrupt school and career plans, and limit what people can expect for themselves in heartbreaking ways. The suffering almost always extends beyond the person with the disorder. Family life can be thrown into chaos. Divorce is a common consequence. Those closest to the sufferer can themselves develop mental disorders like anxiety or PTSD (posttraumatic stress disorder); they can simply burn out and abandon their suffering friend or family member to preserve their own health. At least half of the people in homeless shelters suffer from mental or substance use disorders.[13] The same is true of prisons.[14] Mental disorders can contribute to violence—not just the school shooters who make it into the headlines, but also domestic violence. Mental disorders can result in such extreme hopelessness that people take their own lives.

For the majority of people, however, mental disorders don't show themselves in dramatic and easily visible ways. Instead, people suffer alone in silence. They are ashamed. They don't know what to do about their symptoms. Oftentimes, they don't even know they have an illness. They don't think of their symptoms as "symptoms"; they think their suffering is just a natural part of existence. They might believe that they are weak or inferior to others. They may think that they just need to make the most of the life they were given. They experience their distress, their symptoms, as an integral part of themselves or their life experiences.

For example, imagine a woman we'll call Mary. Her father was an alcoholic and verbally and physically abusive. He found fault with seemingly everything that she did, and she grew to believe that she was stupid and had few redeeming qualities. She didn't talk to people about her father's rages; she assumed this would only cause more trouble and incite further punishment from him. By high school, she was depressed, isolated, and saw little hope for her future. This continued into her adulthood. Mary had trouble sleeping, had flashbacks of her father yelling at her, and startled easily at loud noises. It did not occur to her that any of this might constitute a "disorder," much less be treatable. I see patients like Mary who have suffered like this for years before something brings them to treatment. Many, many people like Mary never seek treatment at all.

What About Treatment?

Treatment for mental disorders is vitally important. It can reduce suffering. It can prevent disability. It can restore people's dreams and potential. Treatment can save lives. In fact, it does. Many people benefit tremendously from present-day mental health treatments. Patients overcome addiction, find relief from psychotic episodes, learn to manage anxiety, recover from eating disorders—these victories are real and significant. The treatments we have work. Unfortunately, they just don't work all the time or for everyone.

Let's look at a success story first.

John was a thirty-six-year-old engineer, married with two young children. Life had been pretty good for him . . . until he discovered his wife was having an affair. John wanted to save his marriage, but his wife wanted a different life, and she had decided to leave him. John was devastated and became severely depressed. He was unable to sleep more than two hours at a time. He couldn't stop obsessing about how his life was now ruined. He couldn't focus at work. He felt the only solution was to get his wife to return, but she wasn't interested. He was tormented with guilt about all the ways he thought he'd failed as a husband, as a father, as a person. This went on for three months and showed no signs of improving—if anything, it was getting worse. Finally, John's family encouraged him to see a psychiatrist. He came away with a prescription for an antidepressant and a sleep medication, and he started weekly psychotherapy.

Within days, John was getting more sleep. This helped him to feel less disoriented and overwhelmed, but he was still distraught. Within a month, however, things started to turn around. His mood began to improve. He was able to stop the sleep medication and sleep normally on his own. He was managing to focus less on tortured ruminations and more on things that he could control. He focused on projects at work and around the house, and he made the decision to get into better shape. He started spending more quality time with his two children. He took the steps he'd been avoiding to finalize the divorce. After a few months, he was able to stop psychotherapy. After a year, he tapered off the antidepressant and continued to feel well. He started dating again.

John's story demonstrates the success of modern psychiatry. The combination of medications and psychotherapy alleviated his depression and anxiety and helped him cope with an extraordinarily stressful time in his life. John's was not the only suffering lessened, either. Divorce is difficult for children—in fact, it increases their risk of experiencing mental health challenges of their own. Having a parent with severe depression also increases this risk. Treatment allowed John to be a better, more engaged father. So, helping John feel better was also beneficial to his children. John's workplace benefited, too. While depressed, John still showed up at work every day, but he wasn't able to focus, and he got less done. Successful treatment helped John be a more productive employee.

There are countless stories like John's, and it is easy to understand why researchers and clinicians in the mental health field like to tell them. It's important to highlight that treatments can work. It's important to encourage people to seek help, to let them know their suffering can end. And professionals in any field want to focus on their successes. They don't tend to advertise what's not working. Unfortunately, in the mental health field, there is a lot that isn't working. Not everyone gets a positive outcome like John. In fact, most don't.

Depression is one of the most commonly diagnosed and treated mental disorders in the United States. In 2020, an estimated 21 million adults experienced at least one depressive episode, representing 8.4 percent of all US adults. About 66 percent of them received some form of treatment.[15]

So, what happens to all of these people who get treatment for depression? Do they get better—and most importantly, stay better—over the long run?

One study tried to answer this by recruiting a group of people seeking treatment for major depression from five different academic medical centers and following them for twelve years.[16] The study included 431 people, and researchers assessed their symptoms of depression on a weekly basis. What they found was that even with treatment, 90 percent had persistent symptoms. On average, over the twelve-year period, the people in the study had symptoms of depression 59 percent of the time. Their symptoms would fluctuate, sometimes going away but then coming back, even with treatment, even if they took medications every day. In other words, 90 percent of the people were not cured of their depression. They either continued to have low-grade lingering symptoms or they would come in and out of bouts of major depression. Depression was found to be a chronic, but episodic, illness. These researchers found that if people had only one episode of depression, like John, the chances of a full and lasting recovery were greater. However, there weren't many of those people.

This study is not an outlier; it is a reflection of what anyone who has worked in the mental health field for a number of years already knows to be the case. Almost two-thirds of depressed patients don't experience remission—meaning get all the way better, even temporarily—with the first treatment they are offered.[17] As the statistics suggest, many people go on to suffer for years, despite trying treatment after treatment. It's not just a failure of medications, either. Many people try numerous treatments—medications, psychotherapy, group therapy, meditation, positive thinking, stress management, and more. Some even try transcranial magnetic stimulation (TMS) or electroconvulsive therapy (ECT, also known as "shock therapy"). People for whom no treatment seems to be very successful are said to suffer from "treatment-resistant depression," though there are many more people who do find some relief, but not in a complete or durable way. The fact that depression is the leading cause of disability in the world clearly speaks to the lack of effectiveness of our current treatments. What are we missing? Why can't we get most people with depression all the way better and keep them better?

You may be wondering about the outlook for mental disorders other than depression. Sadly, the statistics for many other disorders are even worse. I won't go through the data for every condition, but disorders like OCD (obsessive-compulsive disorder), autism, bipolar disorder, and schizophrenia are all at least as bad as depression in terms of both treatment success and the chronic nature of their afflictions.[18] Many of these patients are told that they have lifelong disorders and that they will need to lower their expectations about what they will be able to achieve in life.

Understandably, many patients are frustrated by the ineffectiveness of mental health treatment. They hear stories like John's and assume they should be cured like he was. They often come to believe that the professionals treating them are incompetent, or that they haven't gotten the correct diagnosis, or that they just haven't found the right pill yet. Unfortunately, these usually aren't the reasons they aren't getting better. For most people, it's simply because our treatments don't work all that well.

Some professionals in the mental health field won't like this assessment or approve of me sharing it this way. They may fear that pessimism about treatment will deter people from seeking help. This is a legitimate concern. It is important that those suffering from mental illness reach out for support from professionals—sometimes this can be enough to keep someone alive through a suicidal crisis. Nonetheless, the data I've presented is accurate; to claim that mental health treatment works for everyone (or even most people), and works completely, is misleading at best. A bigger concern is that these types of claims can serve to further shame and stigmatize those with mental disorders. If people are told that our treatments work, and then those people don't get better, some will blame the treatment or professionals, but others will blame themselves. And it isn't just the patients: If we make these kinds of claims to families, other clinicians, and society at large, what happens when patients don't get better? Do we say that the patient must have a "treatment-resistant" version of the disorder, implying that they have a more severe form of mental illness (which may very well not be true) and possibly adding to the stigma they suffer? Or do we suggest that it is the patient's fault? Is the person not trying hard enough in therapy? Does the person somehow "want" to be sick? Unfortunately, these kinds of implications are all too common from clinicians, family members, friends, and others. And so we are back where we began, with the choice to come clean and say that for most disorders, treatments don't work long-term for the majority of people. This brings with it the risk of discouraging those who need it from seeking treatment in the first place.

Given everything I've outlined in this chapter—that these disorders are common and becoming more common, that they are an enormous burden on society both in terms of economic impact and human suffering, and that our treatments have proved unequal to the task of relieving that burden—it seems clear that mental illness is a global health emergency. We have poured money into research in the hope of shedding light on the problem and uncovering new solutions. In 2019, the National Institutes of Health (NIH) spent $3.2 billion on mental health research. What do we have to show for the research that has been done?

This is what Dr. Tom Insel, the former director of the National Institute of Mental Health (NIMH), had to say in 2017 after leaving the NIMH:

> *I spent thirteen years at NIMH really pushing on the neuroscience and genetics of mental disorders, and when I look back on that I realize that while I think I succeeded at getting lots of really cool papers published by cool scientists at fairly large costs—I think $20 billion—I don't think we moved the needle in reducing suicide, reducing hospitalizations, improving recovery for the tens of millions of people who have mental illness.*[19]

This was brave of Insel to acknowledge. Those in the mental health field know it to be true. So, again, what are we missing?

The fact is, in order to make real progress, we have to be able to answer the question: "What causes mental illness?" And up until now, we have failed.

Chapter 2

What Causes Mental Illness and Why Does It Matter?

Insanity, madness, anxiety, irrational fear, unrelenting depression, addiction, suicide: Mental illness has been described in every human culture on Earth, stretching back from the present to antiquity. While we have seen that it is on the rise, it is far from a new affliction. And yet the question of what causes it continues to perplex us. Ancient scholars, philosophers, and poets, along with modern-day neuroscientists, physicians, and psychologists, have studied this question relentlessly, and come up with no definitive answer.

Many theories have been proposed over the past few millennia. In ancient times, mental illnesses were largely thought to be the result of supernatural forces. Punishment from God was a common belief. Demonic possession also had its heyday, with exorcism being the treatment of choice. While these kinds of views have persisted and resurfaced throughout history, a more scientific attitude emerged almost as soon as illness in general began to be looked at in a natural rather than supernatural light, and the conception of mental illness as a medical disorder was born. The ancient Greek physician Hippocrates was among those who took mental illnesses seriously; he postulated that they might be due to an imbalance of the body's four vital fluids, or "humors." An excess of one of these, black bile, was thought to cause depression—or melancholia; in fact the word "melancholy" comes from the Greek for "black bile." (Interestingly, bodily substances—particularly feces, as it relates to the gut microbiome—are making a comeback in the theory of mental illness. More on that later.) Just as the birth of medicine transformed the way people thought of mental disorders, so, naturally, did the development of the field of psychology. Sigmund Freud famously theorized that mental disorders were due to unconscious desires or conflicts, and he framed the working of the mind in terms of nonphysical entities or forces—the id, the ego, and the superego. Other psychological theories have since been developed, many attempting to explain mental illness more "scientifically" based upon what we know about behavior and neuroscience. Modern cognitive or behavioral theories, for instance, might view an anxiety disorder as the result of internalized patterns of thought, or advocate changing certain behaviors as a way to change mental experiences. While psychological theories are still used in treatment today, most clinicians and researchers do not believe they can explain *all* mental disorders. From the mid-nineteenth century to the present day, there has been increasing evidence that mental illnesses have at least some biological components or influences. Chemical imbalances, brain changes, hormones, inflammation, and immune system problems are all believed to have possible roles in causing mental illness. Nonetheless, some authorities in the field feel that a physical model of mental states is too "reductionistic." They say it reduces the complexity of human behavior, emotion, and experience to chemistry or biology, and that human experience cannot be explained by our mere molecules.

In 1977, Dr. George Engel, an internist and psychiatrist, developed a working model of what causes mental illness that is still widely used today. He called it the *biopsychosocial model.*[1] It asserts that there are (1) biological factors, including genes and hormones; (2) psychological factors, such as upbringing and rigid beliefs; and (3) social factors, like poverty or a lack of friends, that all come together for any given individual to produce a mental illness. Another popular model is the *diathesis–stress model.* "Diathesis" means a biological predisposition to becoming ill, something like genetics or a hormonal imbalance. The stress in this model can be anything in the environment, such as getting fired from a job, drug use, or even an infection, that then pushes the already predisposed person to actually become ill. This model assumes that most people who develop mental disorders were likely to do so at some point or another in their lives—they were just waiting to be triggered. Both of these models present a picture of mental illness that attempts to account for many different factors all contributing to the development of a mental disorder.

Because, in fact, we *have* identified many factors that make people more likely to develop various mental disorders. And today, when we think about what causes mental illness, we often think in terms of these risk factors. They include things like stress, drug and alcohol use, hormonal issues, and a family history of mental illness, among others. The problem is that although we know of many such risk factors, not one of them is present in everyone with a specific disorder, and not one of them is *sufficient,* in and of itself, to cause any specific disorder.

An obvious example is PTSD. This is the disorder that causes people to experience fear, flashbacks, excessive anxiety, and feelings of numbness for months or years after living through a traumatic event. By definition, anyone with PTSD must have been exposed to a traumatic event, but only about 15 percent of people who experience such trauma end up developing PTSD. Even when two people experience the same traumatic event, one person can end up with severe PTSD while the other is spared entirely. In other words, trauma, on its own, does not "cause" PTSD. *OK,* you say, *that's because it results from a combination of risk factors.* Unfortunately, there is no combination of risk factors "guaranteed" to result in PTSD, either. And so it goes with almost every other mental disorder. Sometimes, it seems easy to understand why someone develops mental illness. For example, a woman who had a horrible, abusive childhood, who has a thyroid disorder, and who was just left by her husband of ten years for another woman might develop clinical depression. Most people can understand why she might be depressed because she has many risk factors for developing clinical depression. However, there are other people for whom mental illness seems to come from out of the blue for no reason at all.

Decoding Depression

Let's take a look at one of the most clearly defined and understood mental disorders—major depression. Everyone gets depressed, but not everyone gets major depression. People with major depression feel sad or depressed most of the time, and they may experience fatigue, problems concentrating, and disrupted sleep. This disorder can rob people of all ability to experience pleasure and enjoyment from life, and it may leave them with overwhelming feelings of hopelessness and even thoughts of suicide. There are nine symptoms in all, and to be diagnosed with major depression, a person must experience at least five of them for at least two weeks.

There are many clearly established risk factors for the development of major depression. They include genetics or a family history of depression, stress, the death of a loved one, the breakup of a relationship, conflict at work or school, and physical and sexual abuse. Various hormonal problems also make the list, including low thyroid hormone, high levels of cortisol, and the fluctuations in women's hormones that can contribute to the risk of depression in the postpartum period or around the time of menstruation. In fact, just being a woman doubles the risk of depression compared to being a man. Excessive drug or alcohol use is a risk factor, and even some less obvious prescription medications, like certain antibiotics or blood pressure medications, can increase risk as well. Then there are social issues, like being bullied or teased, having no friends, or simply feeling lonely most of the time. Poverty, malnutrition, and unsafe living environments also increase risk. Sleep disturbance is a big one: either too much or too little sleep puts people at risk for depression. Many physical illnesses make the list of risk factors, including chronic pain, diabetes, heart disease, and rheumatoid arthritis. Cancer is another risk factor—but not necessarily the way you might think. Most people would be stressed by a diagnosis of cancer and assume it's only natural that such a devastating diagnosis might make someone depressed. That does happen for some people. However, some patients get clinically depressed *before they even know* they have cancer. This is a common occurrence with pancreatic cancer in particular—people find themselves depressed for seemingly no good reason, and then a few months later are diagnosed with pancreatic cancer. Virtually all neurological disorders are associated with higher rates of depression, including strokes, multiple sclerosis, Parkinson's disease, Alzheimer's disease, and epilepsy. And interestingly, every other psychiatric disorder puts people at much higher risk for also developing major depression on top of their already existing disorders.

That's . . . a lot of risk factors. And they vary widely. Not just in the kinds of things they are—biological, psychological, and social—but in how big of a role they are believed to play. For instance, while being a woman is a risk factor for major depression, no one would say that being a woman causes major depression. But there are certain factors that contribute more directly, and in fact many competing theories hold one factor or another to be the root cause of the disorder. Unlike those who subscribe to the biopsychosocial model, some professionals believe that major depression has a purely genetic, purely biological, or purely psychological origin—the rest of the risk factors are just window dressing.

One of the most widely known of these single-cause theories is the *chemical imbalance* theory of depression—and in fact, of mental illness. This theory suggests an imbalance in brain chemicals called *neurotransmitters* is the cause of all mental illness. Neurotransmitters are chemicals that relay signals between brain cells. For depression, the most popular belief is that levels of the neurotransmitter serotonin are too low; thus, medications that increase serotonin levels will treat depression. Many of the most commonly prescribed medications for depression belong to a class of antidepressants called selective serotonin reuptake inhibitors, or SSRIs (Prozac, Zoloft, and Paxil are in this category). They often *do* help resolve the symptoms of depression, which supports the theory that maybe a chemical imbalance is the cause. Other classes of medications that affect different neurotransmitter systems can also alleviate depression, so maybe it's not just serotonin—maybe it's various neurotransmitters in different people. Nonetheless, many psychiatrists and researchers believe that at its root, depression always boils down to a chemical imbalance.

However, there are many questions raised by this theory:

- What causes the chemical imbalance in the first place?
- If people are born with this chemical imbalance, why aren't they depressed all the time, starting from birth?
- Why do medications like SSRIs take weeks or months to work? We know they change neurotransmitter levels within hours, so why don't they work immediately?
- If it's a fixed chemical imbalance, why do the symptoms wax and wane, even over short periods of time? Put another way, why do people have good days and bad days, even when taking medications consistently?
- Why do the medications stop working in so many people? Why would the imbalance change, and if it does change, what makes it change?

These questions are begging for answers, not just in relation to the diagnosis of depression, but for all psychiatric diagnoses. Unfortunately, the chemical imbalance theory doesn't provide answers.

Another widely known theory for the cause of major depression is the *learned helplessness* theory. In a nutshell, it holds that when people are unable to change adverse circumstances in their lives, they "learn" that they are helpless. This can apply to something like not being able to find a romantic relationship despite numerous attempts, or more direly, an abused child trying to get his father to stop hitting him. In either case, these people begin to feel powerless, and then they get depressed. Eventually they stop trying to do much at all. Why bother? Some experts assert that the cause of these people's depression is their psychology. They have learned, and believe, that they are helpless. Obviously, getting the abused child out of that environment is of paramount importance. But even years later, that boy may still be depressed. The treatment is often based on cognitive behavioral therapy (CBT), a form of talk therapy that focuses on identifying and changing thoughts, emotions, and behaviors. This therapy is based on the belief that when people are clinically depressed, it is likely because of thoughts that are based not so much on the reality of their current situation as on the helpless mindset developed in the past. The goal is to empower patients to challenge these thoughts and replace them with ones not so dire and hopeless. This will help them feel better and make changes in their lives, which will make them feel less helpless still, and this cycle will reinforce itself. This treatment works, at least for some people, which again supports the theory that this kind of problem might be a cause of depression.

There are many other theories about specific factors believed to be the cause of major depression—biological, psychological, and social. Many have led to the development of specific treatments and interventions that work with real people, at least some of the time. In fact, the theories themselves are often shaped by treatments that are effective for depression, using the logic that if a treatment works, even in some people, then it must be correcting a problem that was causing the illness.

Medications used to treat major depression include those specifically known as "antidepressants," which are commonly sorted into five different classes. These classes act on different neurotransmitters and receptors, including ones for serotonin, dopamine, and norepinephrine. However, antidepressants aren't the only medications used to treat major depression. Others include anxiety medications, mood stabilizers, antipsychotics, stimulants, antiepileptic medications, hormones, vitamins, and a wide variety of supplements, such as St. John's wort. These all work in very different ways, and yet all are used routinely in treating depression, and all have been shown to work for at least some people, some of the time.

Psychotherapy to treat depression also comes in many varieties. Some focus on relationships, others on thoughts and feelings, and others on behaviors; some focus solely on changes in the present, and others on revisiting your past or childhood. Different types of psychotherapy can be very different from each other, yet there is at least some evidence that they can all be helpful to at least some people with depression.

Finally, there are more aggressive treatments, like TMS, ECT, and even surgery, in which parts of the brain are severed or electrodes are implanted to stimulate the brain or the vagus nerve, the main nerve of the parasympathetic nervous system.

That's a lot of different treatments! It's difficult to understand how they can all treat the same set of symptoms. However, not one of these works for all people with depression. Why not? Are there just different causes of major depression in different people that require different treatments? And sadly, as I reviewed in the last chapter, there are millions of people who try treatment after treatment without finding even one that works.

On the flip side, it's important to point out that not everyone who develops major depression even gets treatment—in fact, the majority of sufferers throughout the world do not. Yet major depression will often resolve on its own. The symptoms can come and go, sometimes lasting a few weeks or months and then spontaneously disappearing. What causes some people's symptoms to go away without any treatment? Why, for others, does depression become a chronic and debilitating illness? If we truly understand what causes this disorder, we should be able to answer these questions.

But the picture gets more complicated still. In addition to risk factors or theories about what causes major depression, we have good evidence of physical changes in the body that are *associated* with major depression—that is, they are found more often in those with the disease than those without it. These are changes that have been observed in people who already have the diagnosis, but they may also provide clues about the cause of the disorder.

Inflammation is a big one. We know that when compared to those without depression, people with chronic depression, on average, have higher levels of inflammation, as measured by different biomarkers, such as C-reactive protein and interleukins.[2] At this point, however, we don't know for sure if the inflammation is causing the depression or if the depression is causing the inflammation. And if the inflammation is causing the depression, what starts the inflammation in the first place? Is it one or more of the risk factors that we've discussed so far? Or is it something else altogether that we just haven't discovered yet? As usual, many people have theories—some speculate that it is a chronic infection, or an autoimmune disease, or exposure to a toxin, or a bad diet, or having a "leaky gut," and on and on—but these theories are not answers. What's more, not everyone with chronic depression has higher levels of inflammation, at least not that we can measure. The research showing higher levels of inflammation is based on comparisons between *groups* of people: When looking at a group of people with depression and a group without, the group with depression has more inflammation . . . but not every individual in the depressed group will have higher levels of inflammation than the individuals in the group without depression. In fact, researchers and clinicians have not identified any parameter for measuring inflammation in the body or brain that will consistently separate people who are suffering from depression from those who aren't.

In addition to differences in inflammation levels, we have identified differences in the brains of people with chronic depression. Some people with depression have shrinkage, or *atrophy*, of specific brain regions, and this can progress over time. Because these types of changes are often seen in neurodegenerative disorders, some researchers speculate that depression might be a neurodegenerative disorder as well, or that it could represent the early stages of another neurodegenerative disorder such as Alzheimer's or Parkinson's.[3] Other researchers speculate that these changes might be the result of the increased inflammation associated with depression. We know that inflammation over a prolonged period of time can cause damage to tissue. For example, when a person's knee is inflamed by arthritis, we know that permanent damage can result; the longer the inflammation lasts, the more the damage progresses. Maybe something like that is happening in the brain—the inflammation comes first and causes damage to these brain regions.

Research has also found a number of differences in the way the brains of depressed people function. When comparing functional MRI scans of those with major depression to those without, depressed people seem to have decreased activity in some brain regions and increased activity in others, as well as differences in the way brain regions communicate with each other.[4] However, as with all of these brain changes we've discussed, the studies have only shown relative differences between groups. And once again, we don't know if these changes are the cause of depression or a consequence of it. Could another process be causing both the depression and these brain changes? We just don't know yet.

Finally, let's throw another wrench into the works—the *gut microbiome.* The human digestive system contains trillions of microorganisms, including bacteria, viruses, and fungi. They produce hormones, neurotransmitters, and inflammatory molecules that get released into our gut and then absorbed into our bloodstreams. Research suggests that these microbes play a role in obesity, diabetes, cardiovascular disease, depression, anxiety, autism, and even schizophrenia.[5] But microbiome research is relatively new, and we don't yet know the details of which specific microorganisms might be beneficial and which ones harmful, or in fact if it's about the mere presence or absence of certain organisms at all; it could also be that the key lies in the balance of different types of organisms. More to the point, although some research in mice has found changes in depressive symptoms mediated through changes in the gut microbiome, we don't yet know how to use this information to effectively treat depression or most other disorders.[6]

So, that's a whirlwind tour through major depression—many of the risk factors, an overview of some theories about what causes it and the treatments aimed at those causes, and some of the biological and brain changes that are seen in people with the disorder. So, given all that, how do we answer the question: "What causes major depression?"

This is why the *biopsychosocial model* makes sense—there are biological things, psychological things, and social factors that may come together differently in different people to result in major depression. Put another way, there are different causes in different people. Some researchers and clinicians claim that there must be different *types* of depression—maybe one type that's caused by social stressors and another type that's caused by biological factors. Maybe there are dozens of different types of depression—all caused by these various risk factors. Maybe certain factors are responsible for certain symptoms, and if we did a better job of clustering these symptoms, we could identify these types and get a better handle on these causes. Unfortunately, this doesn't appear to be the answer. Clinicians and researchers have struggled with this for decades, and the same set of symptoms continues to come up over and over again across categories, regardless of the risk factors or perceived causes of the depression, whether they are biological, psychological, social, or some combination. The same constellation of symptoms has been found in countless people, in countless varieties of circumstance. In fact, the symptoms of major depression have been described in the Bible, historic texts, literature, poetry, and medical records dating back to Hippocrates. So what causes it? There must be an answer—one that ties together all the facts about the different risk factors, the treatments that work, and the brain and body changes that we see time and again.

Is it possible that there are different processes leading to the same set of symptoms in different people, completely independent of each other? Well, it's possible, but it would be highly unlikely. You may have heard of *Occam's razor*—it's a general rule or guideline, also known as the *law of parsimony.* It is usually held to mean that the simplest, most unifying explanation is most likely to be correct. For example, all things being equal, if a patient comes in with a high fever, sore neck, and headache, it is less likely that the patient has a headache due to a brain hemorrhage *and* a sore neck due to a pinched nerve *and* a fever due to an infection than it is that the patient has meningitis—one diagnosis that explains all three signs and symptoms. In short, when faced with a situation like the one we have outlined for major depression, a unifying theory that can connect all the evidence in a logical and plausible way is most likely to be correct. Before we get too far down the road toward the answer, though, it is worth considering why it matters in the first place.

Why the Cause Matters—Treating Symptoms Versus Disorders

When diagnosing someone with a disorder, we rely on *signs* and *symptoms*. People often use the term "symptoms" as a catchall, but the difference between signs and symptoms is a crucial one. *Signs* are objective indicators of an illness that can be observed or measured by someone else. Signs can include things such as a seizure, a blood pressure measurement, a laboratory value, or an abnormality seen on a brain scan. *Symptoms* are subjective experiences that a patient must tell someone about. Symptoms can include things such as moods, thoughts, or experiences of pain or numbness. There are very few signs in psychiatry. Instead, most of our diagnoses are based on symptoms, such as irritability, anxiety, fear, depression, abnormal thoughts or perceptions, and impaired memory. Mental disorders can also include things that seem more "physical" than "mental," like sleep disturbances, slowed movements, fatigue, and hyperactivity. Some of these can be observed, but clinicians often rely on patients to tell them about these, too, putting them into the category of symptoms as opposed to signs. Unfortunately, there are no laboratory tests, brain scans, or other objective tests that can accurately diagnose any mental disorder.

Psychiatric diagnoses are all based on the concept of syndromes. A *syndrome* is a cluster of signs and symptoms that commonly occur together, with a cause that is not yet known. One medical example that began in the 1980s was the syndrome of unusual infections and rare cancers that we called AIDS—acquired immunodeficiency syndrome. Before we knew that it was due to a virus, it was a syndrome. *In psychiatry, every diagnosis is a syndrome.* This is inherent to the definition of a psychiatric disorder. When mental symptoms are caused by a medical or neurological condition, that alone excludes classifying them as a psychiatric disorder. Neurological illnesses, cancers, infections, and autoimmune diseases can all affect the brain. When people with these conditions have mental symptoms, they are not necessarily diagnosed with a psychiatric disorder. If a patient comes in suffering from irritability, depression, and memory loss, and further evaluation reveals that these symptoms are the result of an infection or cancer, they are diagnosed with that condition and treated by a medical specialist outside of psychiatry, even if their mental symptoms are indistinguishable from those of a patient who "just" has depression. Psychiatrists and other mental health professionals are left with everyone else—the ones for whom we don't know the exact cause.

This is at the heart of the difficulty we've had in making progress in mental health care. Without a clear cause, we end up treating symptoms as opposed to disorders.

Some treatments are designed to attack the root cause of an illness. The best example is an infectious disease. A bacterial infection can cause many signs and symptoms—fever, changes in blood cell counts, chills, pain, cough, and fatigue, to name a few. Definitive treatment for the infection is an antibiotic that eliminates the bacteria from the body. This type of treatment is sometimes referred to as a *disease-modifying treatment*. In this case, the treatment will *cure* the illness; after the course of antibiotics, the person will no longer have the infection. But there is another type of treatment commonly used in the medical field; treatments in this category are known as *symptomatic treatments*. They are designed to reduce symptoms, which can help people feel better, but they don't directly change the course of the illness. For example, people with bacterial infections are commonly given symptomatic treatments like Tylenol to reduce fever. Symptomatic treatments can reduce suffering and allow people to work and function normally, but they are not addressing the root cause. In the end, with or without Tylenol, either the body will fight off the infection on its own, the person will get antibiotic treatment, or the infection will progress and the person will die. Tylenol won't really make much of a difference in which of these outcomes occur.

In the mental health field, the reality is that most of our treatments are symptomatic. For most people, psychiatric medications, ECT, and TMS are usually symptomatic treatments. They don't appear to address the root cause of the illness. For some, they can markedly reduce symptoms. In others, they can put the illness into remission, meaning that all the symptoms get completely better. There are people, like John, who can use antidepressants or other medications for a year or two and then live happily ever after without them. Does this mean that the medications were disease modifying? In some cases, like John's, it's possible they are. However, given the extremely high rates of continuing symptoms and relapses in most people with mental disorders, our treatments don't appear to be modifying the diseases themselves.

As for psychotherapy and social interventions, there are those who believe that these treatments *are* addressing root causes. In some cases, this makes sense. For example, if a woman is in a physically abusive relationship and is clinically depressed as a result, helping her leave that relationship and build a new and better life may resolve her depression. Many psychotherapists would argue that the root cause of the woman's depression was being in an abusive relationship. However, we know that because she has gone through the experience of an abusive relationship and developed clinical depression, she is now at increased risk for developing depression again at some point in the future, even if she is never in another abusive relationship. Given this fact, it seems there may be more to the depression than just the abuse, and that treating this factor alone did not definitively address the root cause. What causes her to remain at higher risk of developing depression in the future? If we truly understand what causes mental illness, we should be able to answer this question.

People in the mental health field often use circular logic to support their theories about what causes mental illness. For example, they may claim that if something works to relieve symptoms, it must have been the cause to begin with. The fact that the woman in the example above found relief when her situation changed is used as evidence that this situation was the root cause of her clinical depression. The fact that many psychiatric medications help relieve the symptoms of mental illness is used as evidence that the root cause of mental illness must be a chemical imbalance. As logical as this may seem, it isn't always true.

Here's an example to illustrate some of the flaws in this line of reasoning. Let's go back to an infection that is causing a fever. If we didn't know anything about infections or the causes of fever, and we were trying to figure it all out, we might do brain scans on people with fevers, looking for clues. Guess what we would see? We would see that the hypothalamus is overactive—that's the part of the brain that controls the fever response. If we already knew that Tylenol works to reduce fevers, we might then do scans to research how Tylenol affects the brain. Lo and behold, we would see that Tylenol decreases this excessive activity in the hypothalamus! Based on this, we might logically conclude that the cause of the fever is a brain disorder involving the hypothalamus. We would have proof that a fevered patient's brain activity is abnormal and that Tylenol works to reduce that abnormal activity. But it would be deeply misguided to conclude that we'd identified the fever's *cause*. What we really did was identify a part of the brain that is involved in the production of a fever, and prove that a treatment that reduces fever affects that part of the brain as well. But Tylenol doesn't treat infections. Reducing fever with that treatment wouldn't change the course of the illness. Our fever and Tylenol brain scans only identified one aspect of the body's response to an infection. We would better understand one *symptom* of the illness, or one part of its mechanism. It's useful information, but it doesn't help us understand the root cause of the fever at all—an infection.

Correlations, Causes, and Common Pathways

To answer the question of what causes mental illness, it is important to think a little about how we go about asking such a question, and the tools and principles we use to explore it. When medical researchers are doing their detective work to determine what causes an illness, they often study groups of people with and without the illness to look for *correlations*. A correlation is a relationship or connection between two things, or variables. If two variables are correlated, it *might* imply a cause-and-effect relationship, which is ultimately what the researchers are looking for. There are many types of studies designed to look for correlations. Researchers might perform brain scans on groups of people with and without depression and look for differences—the association with inflammation mentioned earlier is a correlation, the result of noticing that two variables (depression and inflammation) seemed to occur more often together, implying a relationship. One common type of study is an *epidemiology study,* which assesses variables in large populations of people and looks for correlations that way. For example, researchers might measure people's weight, follow them for ten years, and record how many suffer heart attacks over that ten-year period. They would then look at the rates of heart attacks for different groups of people based on their starting weights to see if weight was correlated with heart attacks. If they found that obese people had higher rates of heart attacks than thin people, they would conclude that there was a correlation between obesity and suffering a heart attack. Notice that I said "correlation." Based on this study alone, they can't say that being obese *causes* heart attacks. This is one of the tricky things about correlational research. People often misinterpret the findings and make assumptions that aren't warranted.

Correlation does not equal causation. Almost everyone has heard this. It means that a correlation does not necessarily tell us anything about cause and effect. Unfortunately, while most people are aware of the principle, they don't apply it when interpreting research. If the study in the example I just discussed were released today, the headlines would likely read "Obesity Proven to Cause Heart Attacks," further perpetuating the incorrect interpretation of studies like this. This may seem like semantics. You might be thinking, "Of course obesity causes heart attacks. What's your point?" Well, as a matter of fact, obesity, in and of itself, does *not* cause heart attacks. It is a strong risk factor for having a heart attack, but it's not a definitive cause. What's the difference? Not all obese people experience heart attacks. If obesity causes heart attacks, all obese people should have them, and probably have them often. Also, there are plenty of people who are stricken by heart attacks who are not obese. If obesity is the cause of heart attacks, why would a thin person ever have one? Clearly, there must be more to heart attacks than obesity. So, what *does* cause a heart attack? The correct answer might be something along the lines of "the heart's arteries develop atherosclerosis (thickening or hardening) and, at some point, become obstructed, resulting in some of the heart muscle dying or being damaged due to a lack of blood flow." What causes those things to happen? That's where obesity comes in as a risk factor, but other risk factors also contribute to this process, such as genetics, cholesterol and lipid levels, blood pressure, a lack of exercise, stress, poor sleep, and smoking. There is a *cascade of events* that leads to a heart attack; this cascade of events occurs over years. Understanding the entire cascade of events is important, as it offers numerous opportunities to intervene with different treatments. If we assume the cause is obesity and focus all our treatments on this risk factor, we will fail to prevent heart attacks in many people. How we define the *cause* of illness matters. Everyone likes simple answers. The way I just defined the cause of a heart attack is a complicated answer. As you will learn, this is also the case when answering the question: "What causes mental illness?"

Correlations, or relationships between two variables, can exist for several reasons. The most common interpretations are *cause* or *consequence*: One variable causes the other or is a consequence of the other. In other words, if A and B are correlated, it might be because there is a cause-and-effect relationship, where A causes B or B causes A. However, there is another possibility—one that some people have trouble understanding. Correlations can also reveal a *common pathway* or, sometimes, a *common root cause.*

Let's assume we don't yet know anything about the cold virus. All we know is that lots of people are showing up at the doctor with runny noses and sore throats. Some people also have other symptoms, such as headaches or fatigue, along with their runny noses and sore throats. Some have only one or the other—a runny nose *or* a sore throat—but many, many people have both. Researchers notice that there is a correlation between runny noses and sore throats. Since there is a correlation, there must be a relationship. But what relationship? Is it cause and effect? If so, which causes which? Many people seem to get their sore throats first and then develop runny noses, but not all—in fact, sometimes it is the opposite. So does the sore throat come first and cause the runny nose? Is it the other way around?

Or are they both simply consequences of some unidentified illness that can cause both symptoms, and maybe even others?

Even though this is a straightforward example of an infection with the cold virus, at one point in time this all needed puzzling out. One source of confusion might have been people with allergies developing runny noses and sore throats when pollen levels were high. These people would present with the same, or similar, symptoms, but from a different root cause—allergies instead of the cold virus. The researchers would have to work hard to tease these two groups apart, trying to sort patients in various ways. At the end of the day, the symptoms in question would likely be indistinguishable: a runny nose is a runny nose, whether it's from allergies or a cold. Researchers might have more luck by noticing things like seasonal patterns, or that some people seem to spread their symptoms to others (those with the cold virus) while others don't (allergies). Looking for and combining patterns would give the researchers important clues about how to distinguish the two groups. In the end, they would have to address this important question: Are the runny noses and sore throats in these two different groups of people related in some way? They are, after all, the same symptoms. Why?

The answer is that they share a common pathway—inflammation. Inflammation is part of the body's process of healing tissue and/or fighting off an attack, and it occurs whenever the immune system is activated. Whether the body is mounting a defense against a cold virus or an allergen, inflammation is causing the runny noses and sore throats. Inflammation is the common pathway or process that produces the symptoms for both groups of patients, but that pathway is downstream from the root cause. To get to the root causes, the researchers will need to determine what's causing the inflammation.

Another way researchers might try to understand the symptoms of runny noses and sore throats and their causes might be to look at them separately. Not everyone has both symptoms, and some people with both have mostly one or the other. Our researchers might sort people into a group of those with primarily or only runny noses and another of those with primarily or only sore throats. This could make sense. After all, noses are different than throats. The treatments for the two symptoms are also different. Tylenol might help soothe the pain of sore throats, but it wouldn't help runny noses. The most effective treatments for runny noses would primarily be things like pseudoephedrine or phenylephrine, ingredients found in Sudafed and cold and flu medications. There might be some treatments that relieve both symptoms in some patients—for instance, an antihistamine in those with allergies—but Tylenol would help almost all sore throats, and pseudoephedrine almost all runny noses, while neither would affect the other symptom. This stark difference in treatments might support the categorization of people into a runny nose group or a sore throat group. The researchers might label these distinct disorders—runny nose disorder and sore throat disorder.

Given that the treatments for these disorders are so distinct as well, they might instead think of these disorders as they relate to those treatments. The sore throat group might be thought of as having a Tylenol deficiency disorder: sore throats must be due to people not having enough Tylenol in their system, as correcting this deficiency seems to rectify the problem. The other group might be called pseudoephedrine deficiency disorder, as the effectiveness of the treatment clearly indicates a pseudoephedrine imbalance in the body.

As facetious as this seems, this is the very same logic that we use to conclude that depression is due to a serotonin deficiency and psychosis is due to too much dopamine. It makes sense, until you think about it with an example like the cold virus, which we understand well. In this context, it seems ridiculous. Yet that's what we are doing in the mental health field today. We look at the treatments that work and assume that this tells us the story of what's causing the disorders. And the disorders themselves are simply clusters of symptoms we have labeled disorders—the diagnostic labels mean nothing in terms of cause and effect, or what is happening in the body or brain.

Let's return to our hypothetical researchers. These researchers have identified two distinct disorders—runny nose disorder and sore throat disorder. The disorders have different symptoms and different treatments, so the researchers are feeling pretty confident about this classification system. The problem is that, while there are people with just one disorder or the other, *comorbidity* is common—in other words, there are a lot of people who have both disorders. People diagnosed with runny nose disorder commonly also develop sore throat disorder. But it's also true in reverse. Runny noses and sore throats are a good example of a *bidirectional relationship*. It means that if you have either disorder you are at much higher risk for developing the other one. It doesn't matter which one starts first. A bidirectional relationship often means that the two things share some common pathway. For runny noses and sore throats, as I've already discussed, the common pathway is inflammation. Sometimes, in addition to a common pathway, a bidirectional relationship can also imply the same root cause. In this example, we already know that there is a common pathway (inflammation) and different root causes (the cold virus and allergies, among others).

Comorbidity aside, given that the symptoms and treatments are different, researchers and clinicians might advocate for keeping runny nose disorder and sore throat disorder as separate diagnoses. But once someone comes along and identifies the common pathway or a root cause that produces both disorders, this should change. Why? Back to Occam's razor—the law of parsimony. If there is a simpler explanation for something in medicine, that explanation is more likely to be true. In this case, the explanation of the cold virus (a root cause) causing both disorders is much simpler than that of people developing both sore throat disorder (due to a Tylenol deficiency) and runny nose disorder (due to a pseudoephedrine imbalance) at the same time. Identifying that allergies (a different root cause) can cause both disorders would be an equally valid reason to change the medical field's approach to diagnosis based on these symptoms. And of course, identifying the common pathway (inflammation) would be especially helpful, as it would allow the development of more effective treatments—and also explain why the symptoms of two different root-cause disorders, a cold and allergies, can be identical.

But the same root cause can also result in different symptoms in different people . . . especially when preexisting vulnerabilities come into play. The flu is a great example. People infected with this virus usually experience a predictable array of signs and symptoms—fever, muscle aches, lethargy, and so on. Yet even though they all have the same illness, different people can have different symptoms to varying degrees. And in people with preexisting conditions, this difference can be magnified. A healthy twenty-year-old might spend a miserable weekend feeling achy and feverish and then bounce back quickly. On the other hand, a child with preexisting asthma might develop severe airway inflammation and end up in the hospital on a ventilator. A frail eighty-year-old man might experience devastating effects resulting in organ damage and death. These people's suffering stemmed from a single root cause—infection with the flu virus—but that cause produced very different consequences.

At this point, you can probably see both why the question of what causes mental illness matters and why it has remained so persistently difficult to answer. We in the mental health field are working with syndromes defined by symptoms and symptomatic treatments. Right now, we are treating infections with Tylenol. The goal is to understand the physiology of mental disorders, enabling us to develop effective treatments, and ideally, prevent these disorders before they occur.

Causation is proving that one thing causes another. Correlational studies on their own simply cannot do this. They can *suggest* causation, or at least provide clues, but proving causation requires something more. One type of study that *can* prove causation is called a *randomized controlled trial*. For example, to prove that the cold virus can cause a runny nose, researchers could take a group of people who are not sick, expose half of them to the cold virus by squirting the virus up their noses and the other half to a placebo (by squirting plain water up their noses). They could then record the number of people in each group who develop runny noses over the following five days. If the cold virus causes runny noses, the group exposed to the cold virus should have a much higher rate of runny noses than the placebo group. In fact, these studies have been done, and this is true.

One of the challenges in proving causation of a serious or life-threatening disorder in humans is that randomized controlled trials are unethical. Thus, even if we do have a plausible theory of what causes cancer, or mental illness, it will be unethical to expose people to this cause to test the theory definitively. So, what can be done in situations like this? Researchers are sometimes allowed to do equivalent experiments on animals. In the mental health field, this can play a role, but it has some limitations given the nature of mental disorders. Another alternative would be to develop a scientific theory of what might be happening in the body or brain from beginning to end—the cascade of events that leads to mental illness, like the cascade of events we talked about earlier as leading to a heart attack. Once established, researchers can then study people who have already been exposed to different risk factors and look for evidence of this cascade of events occurring in them. As you will learn, all of this research has already taken place; the evidence has been gathered. It's just that no one has put it all together. That's what this book does.

Chapter 3

Searching for a Common Pathway

One of the challenges in determining what causes mental illness lies in defining what constitutes a mental illness in the first place. Dictionaries and reference books differ in their exact wording, but a good all-purpose version might be this: *A mental illness involves changes or abnormalities in emotions, cognition, motivation, and/or behaviors resulting in distress or problems functioning in life.* Context matters, though. One of the tricky things about defining mental illness is that many—even most—of the symptoms are considered "normal" in at least some circumstances.

We all have emotions, for example, both pleasant and unpleasant ones. We might feel anxious when we face challenging or threatening situations. We might feel depressed when we experience a significant loss, such as the death of a loved one. Even something like paranoia can have an appropriate time and place. Have you ever watched a scary movie—one that truly terrified you? If so, you were likely a bit paranoid afterward. Some people look in their closets before they go to bed after watching such a movie. Or they hear sounds outside and feel terrified, imagining it's a scenario from the movie. All of this is normal. However, at some point, intense unpleasant feelings and states should diminish, allowing you to go on with your life as before. Therefore, it's important that any definition of mental illness somehow account for context, duration, and appropriateness.

For an example of what I mean, consider "shyness." Are people allowed to be shy? Is that normal? Most would say yes. So, at what point does shyness turn into an anxiety disorder such as social phobia? Drawing these lines is a matter of some debate in the field. One of the most notable controversies concerns depression—specifically whether, in some situations, these symptoms are "normal" and not an illness.

The *Diagnostic and Statistical Manual of Mental Disorders*, or DSM, is the "bible" of psychiatry. It defines all the different diagnoses, their diagnostic criteria, and provides some relevant information and statistics. The current version, updated in 2022, is known as "DSM-5-TR." In DSM-IV, the diagnostic criteria for depression included a caveat called the *bereavement exception*.[1] It suggested that if someone had symptoms of depression in the context of the loss of a loved one, clinicians should hold off on diagnosing depression. A professional might certainly offer support in the form of talk therapy, but prescribing medications wasn't necessarily appropriate. The exception had limits—among them, the depression should not last more than two months and should not produce suicidal thoughts or psychotic symptoms. In DSM-5, however, the exception was removed altogether. This served to encourage clinicians to diagnose depression even in the context of stressful life events like the loss of a loved one. Many clinicians and researchers felt that the American Psychiatric Association (which produces the DSM) had gone too far in pathologizing experiences like grief. On the other hand, supporters of the exemption's removal cited research showing that antidepressants can decrease symptoms of depression even in the context of grief. These advocates felt that not diagnosing the problem and offering medication treatment might be unnecessarily cruel.[2]

Despite such controversies, there are many situations that seem clear-cut. When someone has crippling hallucinations and delusions, or suffers from overwhelming fear and anxiety every time they leave their home, or can't get out of bed for weeks at a time due to severe depression, most of us would agree that this constitutes a mental illness. The "unusual" or "inappropriate" nature or degree of their symptoms, the intensity of their distress, and their inability to function all suggest a serious problem that merits the diagnosis of a disorder.

The premise of the DSM, both the current and prior versions, is that there are distinct mental disorders with clear criteria that can be used to separate them from one another. In some cases, these distinctions are obvious. Schizophrenia is very different from an anxiety disorder. Dementia is different from ADHD. These distinctions are supposed to help guide treatment, predict what will happen to people with specific diagnoses (their prognosis), serve as a tool for clinicians and researchers to communicate more effectively with each other, and so on.

The diagnoses in the DSM are given tremendous importance. They are required for clinical care and reimbursement by insurance companies. They are almost always needed to get research funding given that most mental illness studies focus on just one disorder at a time. And they are critically important to the development and dissemination of treatments as well, since to get FDA approval for a drug, pharmaceutical companies must conduct large clinical trials of specific medications for specific disorders and show a benefit. Even interventions such as psychotherapy are normally studied in clinical trials designed around one specific diagnosis. So, in many ways, the mental health field revolves entirely around these diagnostic labels.

The field, however, has been plagued by debates over how to diagnose different mental disorders, especially since (as we discussed in the previous chapter) there are no objective tests to definitively diagnose any mental disorder. Instead, we use checklists of symptoms and criteria. We ask patients and family members what they are feeling, witnessing, and experiencing; we investigate, cross-reference, and explore; and then we make a diagnosis based on the best match or matches.

In some situations, these diagnostic labels are extremely useful. Remember John who developed major depression? His diagnosis helped inform his treatment, and the treatment worked. John got better—all the way better. After a year of doing well, he was able to stop treatment and remain well. The diagnostic criteria allowed John's psychiatrist to recognize the disorder, understand different treatment options, choose ones that were likely to work, and then discontinue the treatments after a defined period of time. Unfortunately, it's not so simple—or successful—for others.

Sorting Out the Similarities

One of the challenges in the mental health field is that no two people with mental disorders are completely alike, even when they get diagnosed with the same disorder. There are two primary reasons for this—*heterogeneity* and *comorbidity*.

Heterogeneity refers to the fact that people diagnosed with the same disorder can have different symptoms, severity of symptoms, levels of impact on their ability to function, and courses of illness. Not one of the diagnoses requires that all criteria be met. Instead, it's a minimum number—for example, a major depression diagnosis requires at least five of the nine criteria. This makes for a lot of variability. One person with major depression can have depressed mood, excessive sleep, problems concentrating, low energy, and be eating much more than usual, resulting in weight gain. Another person with this diagnosis might be unable to sleep more than three hours, have lost their appetite and shed twenty pounds, and along with depressed mood and low energy be thinking about suicide. These patients have very different symptoms that require differing approaches to treatment. One is thinking about harming himself while the other isn't. One can't sleep, so might benefit from a sleeping pill, while the other one is sleeping too much. Despite these striking differences, both might benefit from an antidepressant or psychotherapy.

Dr. Alan Schatzberg is a prominent depression researcher and a professor of psychiatry and behavioral sciences at Stanford University who has called for rethinking the diagnostic criteria for major depression.[3] Those in this field are frustrated with the lack of understanding of this common illness and continued poor treatment outcomes—as I mentioned earlier, the likelihood of a full and complete remission of major depression symptoms with the first antidepressant a patient tries is only about 30 to 40 percent. Schatzberg notes that some symptoms that commonly occur in people diagnosed with major depression are not included in the core diagnostic criteria. For example, anxiety is a common symptom in many people with depression, but it's not among the nine in the DSM. The same goes for irritability, which is experienced by about 40 to 50 percent of people with depression.[4] Pain is common, too, with physical pain present in about 50 percent of people with major depression compared to only about 15 percent of the general population.[5] Are our treatment outcomes so poor because we're missing or failing to include the treatment of other diagnostic symptoms?

It's not just depression that causes so much confusion and debate. There's tremendous heterogeneity in all the psychiatric diagnoses. Sometimes, the differences are stark and dramatic. Some people diagnosed with OCD are still able to work and function normally in life, while others are completely disabled by their symptoms. People diagnosed with autism-spectrum disorder can be wildly different from each other. There are high-functioning billionaire businesspeople with this diagnosis, while others live in group homes unable to care for themselves. So, are these singular diagnoses really the same disorders? Or are they all simply on a spectrum, with some people having severe forms of the illness while others have a mild form? Unfortunately, the complexities don't end there.

Comorbidity is the other big factor that accounts for differences in people with the same diagnosis. About half of the people diagnosed with any mental disorder have more than one.[6] We talked about comorbidity a bit in the last chapter: Remember my discussion of runny nose disorder and sore throat disorder? While some people had one or the other, many had both. A similar example in the mental health field is that of depression and anxiety. The majority of people diagnosed with major depression also have anxiety, and most people diagnosed with anxiety disorders also have major depression. For example, in a survey of over nine thousand US households,

68 percent of people with major depression also met criteria for an anxiety disorder at some point in their lives, and several studies have found that one-half to two-thirds of adults with anxiety disorders also meet criteria for major depression.[7] Antidepressants are commonly used to treat both depression and anxiety disorders, while anti-anxiety medications are commonly used to treat people with both anxiety disorders and depression. So, where the diagnoses often overlap and the treatments are sometimes identical, are they really different disorders? Is it possible that they are simply different symptoms of the same disorder? Could anxiety and depression—like runny noses and sore throats—share a common pathway?

Finally, diagnoses can change over time. Symptoms can come and go and morph into very different mental disorders, further complicating treatment and diagnosis, and bedeviling the quest to investigate the nature and cause of these disorders.

Let's look at an example.

Mike is a forty-three-year-old man with a chronic, disabling mental disorder. But which one? When he was a child, he was diagnosed with ADHD and started taking stimulant medications. They helped somewhat, but school remained difficult. He was often bullied and teased. He reported a lot of anxiety around these social stressors and received psychotherapy for social anxiety disorder. Some clinicians raised the possibility of Asperger's syndrome, then a diagnosis of its own on the autism spectrum, but they didn't officially make this diagnosis. By adolescence, he'd developed symptoms of major depression—not surprising given both his academic and social stressors. He was started on an antidepressant, which helped a little. Within a few months, however, Mike began to develop symptoms of mania and was quickly diagnosed with bipolar disorder. He had hallucinations and delusions, and he was given medication targeting both his psychotic and mood symptoms. He was hospitalized several times. Over the next year, when his psychotic symptoms persisted and failed to respond to treatment, his diagnosis was changed to schizoaffective disorder. Also during this time,

Mike began to develop obsessions and compulsions, and he was diagnosed with OCD as well. Over the following several years, on top of his continuing psychiatric symptoms, he began to smoke cigarettes and use recreational drugs. Eventually, he became chronically addicted to opioids.

So, what is Mike's diagnosis? According to DSM-5, he currently can be diagnosed with schizoaffective disorder, opioid use disorder, nicotine use disorder, OCD, and social anxiety disorder. But in the past, he also had ADHD, major depression, bipolar disorder, and possibly even Asperger's syndrome. You might make a case that the major depression diagnosis was a mistake—many people with bipolar disorder are diagnosed with depression before they have a first manic episode that clarifies the diagnostic picture. The same might be argued about the change of diagnosis from bipolar to schizoaffective disorder. But even if you remove one or two of these, you are left with a lengthy list of what are all different disorders—supposedly with different causes and certainly different treatments. Yet Mike has just one brain. Are we to believe that he's an extraordinarily unlucky individual who developed some half dozen separate and distinct disorders?

While Mike's story is extreme, having more than one diagnosis is common, as are changes in symptoms and diagnoses. Having problems with addiction is also common in people with mental disorders. Stories like Mike's raise serious questions about the validity of our diagnostic labels. If the diagnoses listed in DSM-5 are really separate and distinct disorders, why do so many people have more than one of them? Why do they change over the course of a lifetime? Do some psychiatric disorders lead to others? If so, which ones come first, and what exactly happens to make them cause other disorders? Alternatively, are some just different symptoms or phases of the same underlying problem? Is it like runny nose disorder and sore throat disorder—two seemingly different disorders that respond to different treatments but share the common pathway of inflammation? Is there a common pathway for mental disorders, even ones that appear to be profoundly different from each other?

Taking a Deeper Look

Researchers have been trying for decades to figure out what makes individual disorders different from each other at a biological level. Interestingly, they don't yet have any clear answers. In fact, as I'm about to share with you, the research to date suggests that different disorders might not actually be all that different from each other, even though symptoms can vary widely.

Let's look at three of the psychotic disorders—schizophrenia, schizoaffective disorder, and bipolar disorder.

The primary feature of a schizophrenia diagnosis is chronic psychotic symptoms, like hallucinations or paranoia. A diagnosis of bipolar disorder is given to people who primarily have mood symptoms—manic and depressive episodes. However, people with bipolar disorder also commonly have psychotic symptoms when they get manic, and even sometimes when they get depressed, but these psychotic symptoms go away after the mood symptoms improve. Schizoaffective disorder is a diagnosis that includes features of both schizophrenia and bipolar disorder, including chronic psychotic symptoms and prominent mood symptoms. Most people consider these disorders unequivocally "real." Many in the field hold these disorders apart from things like depression and anxiety, sometimes calling these the "biological" disorders. So, what do we know about them? What makes them different from each other?

A lot of money has been spent researching this question. The NIMH funded a multisite study called the Bipolar Schizophrenia Network on Intermediate Phenotypes (B-SNIP). This study included more than 2,400 people with schizophrenia, schizoaffective disorder, or bipolar disorder; their first-degree relatives; and people without these disorders (normal controls). The researchers examined key biological and behavioral measures, examining brain scans, genetic testing, EEGs, blood parameters, inflammation levels, and performance on a variety of cognitive tests. They found that people with the disorders were different from the normal controls, but they couldn't tell any of the diagnostic groups apart from each other. In other words, there were abnormalities in the brains and bodies of people with these disorders, but no significant differences at all between those with bipolar disorder, those with schizoaffective disorder, or those with schizophrenia. If they are truly different disorders, how can that be?

On the one hand, when we consider more information, maybe these findings aren't so surprising after all. First, although the diagnosis of schizophrenia isn't supposed to include prominent mood symptoms, the reality is that one of the common features of schizophrenia is a group of symptoms called *negative symptoms*. These include blunting of facial expressions, severely reduced speech and thought, losing interest in life (apathy), getting no pleasure from life or activities (anhedonia), reduced drive to interact with others, loss of motivation, and inattention to hygiene. You might notice significant overlap with the symptoms of depression. Interestingly, DSM-5 specifically cautions clinicians against making the diagnosis of major depression in people with schizophrenia, even though many of these negative symptoms are the same symptoms found in depression. Instead, clinicians are encouraged to diagnose a schizophrenia spectrum disorder. The implication is that even though the symptoms might overlap, we shouldn't call them the same thing. Why not? Is there science to support that recommendation? In reality, the DSM-5 acknowledges in its introductory remarks that we don't know what causes any of the psychiatric diagnoses. Therefore, if people are having the same symptoms, how can we say they are not caused by the same process?

The treatments for these disorders overlap as well—more than you might think. Mood stabilizers, such as lithium, Depakote, and Lamictal, are commonly used in bipolar disorder and have approval from the FDA for such use. However, about 34 percent of people diagnosed with schizophrenia are also prescribed these same mood stabilizers, even though by definition, those with this diagnosis aren't supposed to have significant mood symptoms.[8] Antidepressants are also commonly used in both bipolar disorder and schizophrenia. Studies show that almost all bipolar patients receive an antidepressant at some point in their illness for their depressive episodes, and about 40 percent of patients diagnosed with schizophrenia do, too.[9]

And then there are the antipsychotic medications. These are used for schizophrenia, bipolar disorder, and schizoaffective disorder, and are prescribed to treat *all* the symptoms of these disorders, not just the psychotic ones. The FDA has even approved many of these medications both as "antipsychotics" *and* as "mood stabilizers" for the treatment of bipolar disorder.

On the other hand, while all of this suggests quite a bit of overlap between bipolar disorder, schizoaffective disorder, and schizophrenia, it is also true that the *symptoms* of bipolar disorder and schizophrenia can be dramatically different. Many people with bipolar disorder never have psychotic symptoms. Many are never hospitalized, and many function quite well in life. Meanwhile, almost all people with schizophrenia will experience severe impairment in functioning, with the majority qualifying as disabled.[10] That's not to say that there aren't high-functioning schizophrenics, or that bipolar disorder can't be disabling. In fact, one study that followed 146 people with bipolar disorder for almost thirteen years found that the people were symptomatically ill about 47 percent of the time despite treatment.[11] It's difficult to keep a job when you are ill almost half the time. But there are definite differences in the usual presentation of these diagnoses. Could it be that people with schizophrenia have a more severe form of the same illness, or one less responsive to our current treatments, while people with bipolar disorder may have a milder illness and/or symptoms that respond better to our treatments, resulting in episodes of recovery?

Dr. Bruce Cuthbert, the acting director of the NIMH at the time of the B-SNIP study, suggested, "Just as fever or infection can have many different causes, multiple psychosis-causing disease processes—operating via different biological pathways—can lead to similar symptoms, confounding the search for better care."[12] However, the study failed to find any hallmark biological markers to distinguish the diagnoses. What Cuthbert didn't mention is that we know that fever is itself a symptom, with one clearly defined biological pathway—inflammation that triggers the hypothalamus to increase body temperature. However, there are many things that can trigger the inflammation, such as infections or allergic reactions. Diverse infections can have the same symptoms, through common pathways, even when the infectious agents (bacterial or viral) are different.

It seems very plausible that the symptoms of bipolar disorder, schizoaffective disorder, and schizophrenia all share a common pathway as well.

Sorting Out the Overlaps

I have now suggested that bipolar disorder, schizophrenia, and schizoaffective disorder are possibly the same illness, but on a spectrum of symptoms and with different responses to existing treatments. Earlier in the chapter I suggested that major depression and anxiety disorders might be similarly related and share a common pathway. For many people in the field, neither of these assertions is difficult to grasp or believe. Mental health professionals have struggled with these distinctions for decades and know all too well about the overlap in these disorders and their treatments.

However, the overlap doesn't stop with these conditions.

Symptoms overlap between all kinds of mental diagnoses, not just those you would expect to be related. As I've mentioned, many different disorders, both mental and medical, can lead to psychotic symptoms. In fact, about 10 percent of patients diagnosed with major depression will have psychotic symptoms.[13] Anxiety symptoms are also common in multiple diagnoses. The overall prevalence of anxiety disorders in the general population is quite high to begin with—in any given year, about 19 percent will experience an anxiety disorder. When looking at lifetime prevalence, that number rises to 33 percent, meaning that one out of three people will meet the criteria for an anxiety disorder at some point in their life.[14] The rates in people with depression, bipolar disorder, schizophrenia, and schizoaffective disorder are much higher—about double. Sometimes, we just rationalize these symptoms away: "Wouldn't you be anxious if *you* had schizophrenia?" As appealing and intuitive as this sounds, it's not so simple. There is a strong *bidirectional* relationship between schizophrenia and anxiety disorders. In other words, people who first manifest an anxiety disorder have anywhere from an eight- to thirteen-fold increased risk of developing schizophrenia or schizoaffective disorder.[15] Those are not trivial increases. But why should this be?

In 2005, Dr. Ronald Kessler and colleagues reported the results of the US National Comorbidity Survey Replication, a household survey that included a diagnostic interview of more than nine thousand representative people across the United States.[16] Overall, 26 percent of people surveyed met criteria for a mental disorder in the last twelve months—that's one in four Americans! Of those disorders, 22 percent were serious, 37 percent were moderate, and 40 percent were mild. Anxiety disorders were most common, followed by mood disorders, then impulse control disorders, which include diagnoses like ADHD. Of note, 55 percent of people had only one diagnosis, 22 percent had two diagnoses, and the rest had three or more psychiatric diagnoses. That means almost half the people met criteria for more than one disorder.

Diagnostic overlap is easier to dismiss when we are talking about anxiety disorders, perhaps because anxiety is a mental state we all experience. So let's look at autism spectrum disorder. Most people don't think of autism as a purely "mental" illness, but more as a developmental or neurological disorder that starts early in life. Yet 70 percent of people with autism have at least one other mental disorder and almost 50 percent have two or more.[17] It's also interesting to note that embedded in the criteria of autism spectrum disorder are many of the symptoms of obsessive-compulsive disorder (OCD).

And what happens to people with autism over longer terms? Are they at higher risk for developing additional mental disorders? Again, the answer is often yes. A prominent feature of autism is impairment in social skills, so it stands to reason that a diagnosis of social anxiety disorder could follow if interactions caused anxiety. In such a case, many would assume the autism spectrum disorder came first, and that the social anxiety was an understandable consequence of the autism. However, it's now well documented that autism itself puts people at higher risk for developing every other type of mental disorder.[18] This includes mood disorders, psychotic disorders, behavioral disorders, eating disorders, and substance use disorders. How can this be? Is it just that autism is stressful? We know that stress can put people at risk for all sorts of mental disorders, and having autism is undoubtedly stressful. But as you will learn, the explanation is much more complex than that.

This phenomenon isn't limited to anxiety disorders or autism spectrum disorder, either. Looking at eating disorders, bulimia nervosa occurs in about 1 percent of the population, anorexia nervosa in about 0.6 percent, and binge eating disorder (the newest disorder in the category) in about 3 percent.[19] Many people consider these societal disorders rather than biological brain disorders. Yet overall, 56 percent of people with anorexia, 79 percent of people with binge eating disorder, and 95 percent of people with bulimia have at least one other mental disorder.[20] So here we go again—which one comes first? Do eating disorders cause other mental disorders, or do other disorders cause eating disorders? Both: There is a bidirectional relationship between eating disorders and other mental disorders. Which other disorders, you ask? *All of them.* The same is true of addiction. Again, it's a bidirectional relationship. People with any substance use disorder are at higher risk of developing a mental disorder, and people with mental disorders are at much higher risk of using and abusing addictive substances. Why is that?

I could go on this way, diagnosis by diagnosis, but I won't—an important 2019 study clarifies the bigger picture. In this study, researchers used a Danish health registry to analyze psychiatric diagnoses in almost six million people over seventeen years.[21] What they found was that having *any* mental disorder dramatically increased the chances of that person later developing another mental disorder. *There were strong bidirectional relationships for everything!* Even disorders that most people think are completely unrelated—schizophrenia and eating disorders, intellectual disability and schizophrenia. Mix and match them however you like. The odds ratios across the board in this study were generally between two and thirty. This means that if you were diagnosed with any mental disorder, you were two to thirty times more likely to later be diagnosed with another mental disorder. Which one? Any of them! While some of the very high odds ratios are due to overlap in symptoms between different disorders, the point is that the odds ratios were elevated for *all* disorders in all directions.

What's more, this bidirectional relationship also applied to mental disorders and so-called "organic" mental disorders. "Organic mental disorder" is the term used to refer to symptoms of a mental disorder that are thought to be caused by a medical condition or a medication. We discussed this briefly earlier: For example, if someone with cancer loses their appetite and gets depressed, they often are not diagnosed with major depression. The assumption is that these symptoms are due to the cancer, and not a true "mental" disorder. Yet the evidence of this study now shows that if people develop "mental" symptoms attributed to a medical problem, they are then much more likely to develop a mental disorder in the future—and vice versa. This finding certainly raises the question of whether separating "organic" mental disorders from the rest really makes sense.

All told, this study raised several important questions. Bidirectional relationships, especially ones that are particularly strong in both directions, suggest a common pathway does exist. While the symptoms may differ, perhaps our diagnoses are far more similar than we've long thought.

The Danish study wasn't the first piece of scholarship to suggest all mental disorders might share one common pathway. In 2012, Dr. Benjamin Lahey and colleagues studied the symptoms and prognosis of eleven different mental disorders in thirty thousand people.[22] They looked at "internalizing" versus "externalizing" disorders. Internalizing disorders are ones in which distress is thought to be directed inward, such as depression and anxiety disorders. Externalizing ones are when distress is outwardly directed, such as substance use disorders or antisocial behavior. They found tremendous overlap in these different disorders and raised the possibility of a "general factor" leading to all of them.

In 2018, doctors Avshalom Caspi and Terrie Moffitt took this research further by including *all* mental disorders in a review article, "All for One and One for All: Mental Disorders in One Dimension."[23] They reviewed a tremendous amount of research, including epidemiological studies, brain imaging studies, and studies of known risk factors for mental disorders, such as genetics and childhood trauma. The data was exhaustive, encompassing research on people of different ages, including children, adolescents, and adults, and from many different parts of the world. After examining all this data, they found strong correlations among all mental disorders. When they looked at risk factors for mental disorders, they found that not one risk factor conferred risk only for a specific disorder—instead, each and every risk factor conferred risk for many. For example, one study they examined looked at the genetics of psychiatric disorders.[24] The study evaluated more than three million siblings, hoping to identify which genes conferred risk for depression, anxiety, ADHD, alcoholism, drug abuse, schizophrenia, and schizoaffective disorder. Given that these are all different disorders, one would expect them to have different associated genes. However, the researchers found that most of the genetic variations conferred risk for a broad range of disorders. There were no genes that were specific to only one disorder. Even childhood abuse confers risk for most mental disorders, including PTSD, depression, anxiety, substance use disorders, eating disorders, bipolar disorder, and schizophrenia.

Given the never-ending overlap in correlations among all mental disorders and all of their risk factors, Caspi and Moffitt used a complex mathematical model to analyze these correlations in the hope of making sense of them. This model offered a shocking conclusion. It suggested that there appears to be one common pathway to all mental illnesses. Caspi and Moffitt called it the *p-factor*, in which the *p* stands for general psychopathology. They argued that this factor appears to predict a person's liability to develop a mental disorder, to have more than one disorder, to have a chronic disorder, and it can even predict the severity of symptoms. This p-factor is common to hundreds of different psychiatric symptoms and every psychiatric diagnosis. Subsequent research using different sets of people and different methods confirmed the existence of this p-factor.[25] However, this research was not designed to tell us what the p-factor is. It only suggests that it exists—that there is an unidentified variable that plays a role in all mental disorders.

Our job is to figure out what it could be.

Chapter 4

Could It *All* Be Related?

What if I told you this common pathway that we're searching for might not be limited to mental health conditions?

As we've seen, the medical field currently separates mental disorders from other medical disorders. They are viewed as separate categories that have little to nothing to do with each other.

But there are many medical disorders that commonly co-occur with mental disorders and vice versa. Yes—here we go again with bidirectional relationships: Not only do mental disorders have strong bidirectional relationships with one another, many *metabolic and neurological disorders* also have strong bidirectional relationships with mental disorders. These relationships provide important clues about the nature of the common pathway that will help us solve the puzzle of mental illness.

To explore these relationships, I'm going to focus on three metabolic disorders (obesity, diabetes, cardiovascular disease) and two neurological disorders (Alzheimer's disease and epilepsy). All five of these conditions are commonly associated with mental symptoms like depression, anxiety, insomnia, and even psychosis. On the flip side, people who have mental disorders are at much higher risk of developing these five medical disorders. Clearly not all people with these medical disorders have a mental illness, and not all people with a mental illness develop any of these medical disorders.

When patients with one of these medical conditions do have mental illness symptoms, they're sometimes overlooked as normal reactions to difficult diseases. Those with heart failure are often depressed, which is presented as understandable given the severity of heart failure. And whether people with these conditions who are experiencing mental symptoms get diagnosed with a "mental" disorder is up to clinicians, who have the discretion to attribute these mental symptoms to the "organic" illnesses. In the end, though, the symptoms are the same regardless of what cause they are attributed to. Depression is the same. Anxiety is the same. Paranoia is the same. The treatments are the same as well: antidepressants, anti-anxiety medications, and antipsychotics are all commonly used in people with these "organic" disorders.

Looking at these disorders more closely will illuminate the connections between metabolism, metabolic disorders, and disorders of the brain, whether mental or neurological. They will help us put the final pieces of the puzzle in place.

Metabolic Disorders

Let's start with our three metabolic disorders: obesity, diabetes, and cardiovascular disease. The term "metabolic disorders" actually includes many more disorders, but it most commonly refers to disorders associated with *metabolic syndrome*. This is a syndrome that is diagnosed when people have three or more of the following conditions: increased blood pressure, high blood sugar, excess body fat around the waist, high triglycerides, and low HDL (or "good cholesterol"). People with metabolic syndrome are at increased risk for developing type 2 diabetes, heart attacks, and strokes.

Diabetes

The connection between diabetes and mental illness has been known for more than a century. In 1879, Sir Henry Maudsley wrote, "Diabetes is a disease that often shows itself in families in which insanity prevails." Many mental disorders are associated with higher rates of diabetes. People with schizophrenia are three times more likely to develop diabetes.[1] People diagnosed with depression are 60 percent more likely to develop diabetes.[2]

What about the other way around? Are people with diabetes more likely to develop mental disorders? Yes. Most of the research has focused on depression and diabetes. People with diabetes are two to three times more likely to develop major depression. Furthermore, when they get depressed, the depression lasts four times longer than it does in those without diabetes. At any given time, about one in four people with diabetes has clinically significant depression.[3] What's more, the depression appears to affect blood-glucose levels—diabetics with depression tend to have higher glucose readings than those without depression. However, it's not just depression. One study of 1.3 million adolescents looked at rates of mental illness over the following ten years. Adolescents with diabetes were more likely to suffer from a mood disorder, attempt suicide, visit a psychiatrist, or develop *any* psychiatric disorder.[4]

Obesity

We know that people with mental disorders are more likely to be overweight or obese. One study followed people diagnosed with schizophrenia and bipolar disorder for twenty years. When they were first diagnosed, the majority were not obese. Twenty years later, 62 percent of those with schizophrenia and 50 percent of those with bipolar disorder were obese.[5] The obesity rate at the time for all adults in New York State, where the study was conducted, was 27 percent. Children with autism are 40 percent more likely to be obese.[6] One meta-analysis of 120 studies found that people with serious mental illness were three times more likely to be obese than people without a mental illness.[7]

Many people assume that our treatments are causing this obesity. While there's no doubt that psychiatric medications are associated with weight gain—in fact it's a common side effect of antidepressants and antipsychotics—treatments alone don't provide the entire explanation. For example, one study looked at people with ADHD who were either treated or not treated with medications, and then it assessed their rates of obesity over the ensuing years compared to people without ADHD. They found that *all* people with ADHD, whether treated or not, were more likely to develop obesity. Even though the primary treatment for ADHD is usually a stimulant medication, which generally suppresses appetite, the people with ADHD who were treated with stimulants were still more likely to develop obesity than those without ADHD. Those who didn't take stimulants were even more likely to become obese.[8]

What about people who are obese? Are they more likely to develop a mental disorder? Again, the answer is yes. People who are obese are 25 percent more likely to develop depression or an anxiety disorder and 50 percent more likely to develop bipolar disorder. One study found that weight gain around the time of puberty was associated with a fourfold increase in the risk of depression by age twenty-four.[9] Obesity has been found to affect brain function in ways known to lead to mental disorders as well. For example, people with obesity have been found to have altered connections between brain regions as well as alterations in a region of the brain called the hypothalamus[10] that are common in people with mental disorders.

Cardiovascular Diseases

Cardiovascular diseases—particularly heart attacks and strokes—also have bidirectional relationships with mental disorders. Again looking at depression, we find that 20 percent of people with heart attacks, 33 percent with congestive heart failure, and 31 percent with strokes experience major depression within a year of the event or condition.[11] These rates of occurrence are three to five times higher than those in the US population as a whole.

This seems easy to understand on the surface. Most people would be worried or depressed following a traumatic event like a heart attack or stroke. However, we're witnessing another bidirectional relationship, suggesting this is more than just a psychological reaction.

We know that depression affects the heart. In people who have never suffered a heart attack, experiencing major depression increases the risk of having a future heart attack by 50 to 100 percent.[12] In people who have already suffered a heart attack, being depressed doubles the chances that they'll have another heart attack in the next year.

And it doesn't stop with depression. People diagnosed with schizophrenia and bipolar disorder are 53 percent more likely to develop premature cardiovascular disease.[13] This is even after controlling for risk factors like obesity and diabetes. A thirteen-year study of almost one million veterans found that individuals who were diagnosed with PTSD were twice as likely to have a transient ischemic attack (temporary symptoms of a stroke) and 62 percent more likely to have a stroke.[14]

We've long known that people with serious mental disorders such as schizophrenia, bipolar disorder, and severe chronic depression die at a much younger age than they should. On average, they lose between thirteen and thirty years from their normal lifespans.[15] Recent research from a Danish population database of more than seven million people suggests something more alarming.[16] It's not just the "serious" mental disorders that result in a shortened lifespan. *All* mental disorders—even mild or common ones, like anxiety disorders or ADHD—are associated with shortened lifespans. On average, men with mental disorders lose ten years of life and women lose seven.

What are these people dying from so early? Most think suicide is responsible, but it's not. Although suicide rates are definitely higher in the mentally ill, the early deaths in this group are primarily due to heart attacks, strokes, and diabetes—metabolic disorders. We have just seen that people with mental disorders have much higher rates of these conditions.

Even before they die, we now know that people with chronic mental disorders appear to be aging prematurely. We can see this through a variety of metrics of the aging process. One such metric is the length of telomeres, which are the end caps of chromosomes. They tend to get shorter as people age. Shortened telomeres have been found in people with diseases that you would expect to be associated with aging, such as obesity, cancer, cardiovascular disease, and diabetes. They have also been found to be shorter in people with depression, bipolar disorder, PTSD, and substance use disorders.[17]

Neurological Disorders

Even though both neurological and mental disorders affect the brain and both commonly include "mental" symptoms, they are distinguished based on one thing: Neurological disorders have at least one objective test or pathological finding that can be used in diagnosing the disorder. This can be an abnormality on a brain scan or EEG, or it can be a specific pathological finding in brain tissue or the fluid surrounding the brain. As I've already shared with you, mental disorders have no objective tests that can be used in diagnosis.

Alzheimer's

Alzheimer's disease is the most common form of dementia, which is a group of neurological disorders that impair brain function over time. Common symptoms of all dementias include memory disturbances, changes in personality, and compromised judgment. The hallmark findings in Alzheimer's disease are plaques and tangles in the brain. As people get older, their risk for Alzheimer's goes up exponentially, doubling every five years after sixty-five. By age eighty-five, about 33 percent of all people will have Alzheimer's disease.[18] There are types of early-onset Alzheimer's disease that can be caused by rare genetic mutations or Down's syndrome. However, for everyone else, it's not clear exactly what causes it. Besides age, some of the known risk factors include a family history of the disease, head trauma . . . and metabolic disorders.

Obesity in midlife, diabetes, and heart disease all increase the risk of developing Alzheimer's. So do the *risk factors* for metabolic disorders, like smoking cigarettes, high blood pressure, high cholesterol, and a lack of exercise. Interestingly, one of the genetic risk factors involves a gene variant called APOE4—which codes for an enzyme related to fat and cholesterol metabolism.

Things often considered to be "mental" are also risk factors. Having depression earlier in life doubles a person's chances of developing Alzheimer's disease.[19] Schizophrenia dramatically increases the chances, too—one study of more than eight million people found that if someone with schizophrenia lives to the relatively young age of sixty-six, they are twenty times more likely to be diagnosed with dementia than those without schizophrenia.[20] And remember that large study of the Danish population that found bidirectional relationships among all the different psychiatric disorders? Alzheimer's disease was included under the category of *organic mental disorders,* the label used to classify mental symptoms due to medical conditions, such as delirium and other types of dementia. In that study, *every* psychiatric disorder increased the chances of developing an organic mental disorder—anywhere from a 50 percent increase up to a twentyfold increase. Unfortunately, Alzheimer's disease wasn't separated from the other organic mental illnesses, but the two most common organic mental disorders are delirium and Alzheimer's disease.

The first signs of Alzheimer's disease are usually forgetfulness and "mental" symptoms, such as depression, anxiety, or personality changes. Once Alzheimer's is diagnosed, almost all patients will develop psychiatric symptoms—97 percent in one study.[21] These can include just about any you can think of—anxiety, depression, personality changes, agitation, insomnia, social withdrawal, you name it. About 50 percent of Alzheimer's patients will develop psychotic symptoms like hallucinations and delusions.[22]

So, essentially every psychiatric symptom can emerge with Alzheimer's disease. If that's the case, what's causing these symptoms? Is it the same cause as in people who develop mental symptoms and disorders earlier in life? One thing is certain: this overlap of identical symptoms means we can't truly address the question of what causes mental illness without looking at Alzheimer's disease.

Epilepsy

Epilepsy is a relatively rare brain disorder that also has a bidirectional relationship with mental disorders. Epilepsy can begin at any age, but it most often begins in childhood, affecting about one in 150 children. Sometimes the cause is due to a clearly identified brain abnormality, such as a stroke, brain injury, tumor, or a rare genetic mutation. For most, however, the cause is unknown.

People with epilepsy often have psychiatric symptoms. Sometimes these symptoms lead to a diagnosis of a mental disorder. Other times, however, the symptoms are assumed to be due to the seizures themselves. There is no question that seizures can produce emotions, sensations, or behaviors that are unusual. However, people with epilepsy are also more likely to experience mental symptoms even when they are not seizing.

Twenty to forty percent of children with epilepsy also have an intellectual disability, ADHD, or autism diagnosis.[23] Anxiety disorders are also common in those with epilepsy, occurring at a three- to sixfold higher rate compared to the general population.[24] One study found that 55 percent of people with epilepsy suffered depression, with one-third of all epileptics reporting at least one suicide attempt.[25] Interestingly, the suicide attempts often occurred *before* the epilepsy diagnosis.[26] Other studies have found a sixfold increase in bipolar disorder and a ninefold increase in schizophrenia.[27] The data make clear that psychiatric diagnoses—across the board—are extraordinarily common with epilepsy.

What about the other way around? Indeed, it appears that people with mental disorders are more likely to have epilepsy or experience seizures in general. Anywhere between 6 to 27 percent of children with autism will develop seizures.[28] Signs of epilepsy show up on the EEGs of 16 percent of children with ADHD.[29] Additionally, children who have seizures are two and a half times more likely to *already* be diagnosed with ADHD.[30] Later in life, a diagnosis of major depression increases the chances of having an unprovoked seizure sixfold.[31]

Seizures offer us an important clue on the way to our common pathway, further connecting the dots between metabolic, mental, and neurological disorders: not only is there a relationship between epilepsy and mental illness, there is a relationship between epilepsy and metabolic disorders as well.

We've long known that hypoglycemia (low blood sugar) can cause seizures. This is commonly seen in diabetics—both types 1 and 2. Diabetics can get low blood sugar from either too much medication or not eating enough. However, are people with diabetes more likely to have seizures unrelated to severe hypoglycemia? Yes. Children with type 1 diabetes are three times more likely to develop epilepsy[32]—six times more likely if the diabetes begins before age six.[33] Adults sixty-five and over with type 2 diabetes were found to have a 50 percent higher chance of developing epilepsy.[34]

What about obesity? You might think weight has nothing to do with epilepsy, and yet a large study showed that people who are extremely underweight or overweight were 60 to 70 percent more likely to develop epilepsy than people of a normal weight.[35] That both being over- and underweight are risk factors may be a surprise, but as I'll explain later, both extremes are stressful to metabolism. Additionally, women who are obese during pregnancy are more likely to give birth to children who go on to develop epilepsy, with rates increasing as the mothers get heavier. Women with a BMI greater than 40 have an 82 percent higher risk of having children with epilepsy—almost double the risk in the general population.[36]

Hiding in Plain Sight

So here we are, faced with the curious fact that mental disorders have bidirectional relationships not only with each other, but with these seemingly very different medical disorders as well. Recall that bidirectional relationships suggest a possible common pathway—something in common that is causing or contributing to all these disorders. Is that possible?

Many people think they already know the reason for some of these connections, especially those between metabolic and mental disorders. We've talked about the stigma around mental disorders, but when it comes to metabolic disorders, people are often quick to judge as well. They think those who are obese, diabetic, or have heart attacks are simply not taking good care of themselves. They eat too much, smoke, and/or don't exercise enough. By and large, many believe these conditions arise from negligence—that they're the fault of the diagnosed. Similarly, it seems obvious to many that mental disorders cause people to not take good care of themselves. For instance, depression causes people to lose energy and motivation. When that happens, they sit around all day, watch TV, and eat. They gain weight. They don't exercise. Everyone knows that "stress" contributes to unhealthy habits. Almost by definition, people with mental disorders have more stress than most, or at least it feels that way to them. So, again, people with these stressful symptoms eat poorly and don't exercise enough. It's no wonder that people with mental disorders have higher rates of metabolic disorders. What it really comes down to, in the eyes of many, is simple—these are issues of willpower and discipline.

Here's the conundrum, though. The rates of all these disorders have been skyrocketing over the past fifty years. Obesity, diabetes, cardiovascular disease, and mental disorders. Why is that? Have we had an epidemic of laziness or self-destructive health behaviors in our society? Are people no longer capable of self-discipline? Do they just not care about their health? If you would answer "yes" to these questions, which many would, another question remains: *Why?* What has caused this "epidemic of laziness"?

As we touched on in Chapter One, some might say it's society. The faster pace of everything and the demands of that pace. The stress of modern life. Constant emails to attend to. Social media posts stacking up and vying for our attention. The pull to pick up your cell phone and always be watching, searching, scrolling, or checking. Others might say it's the food supply—artificial ingredients and processed foods.

As it turns out, these *are* likely contributing factors, but are they true causes? How do we get from any one of these "causes" to laziness, apathy, and burnout, which then cause people to overeat and not exercise, which then cause them to develop mental or metabolic disorders? How does that all actually work in the body and brain? And why doesn't everyone subject to these forces end up diabetic and depressed? And where do the connections with neurological disorders, acknowledged as physical brain diseases, fit into all this talk of modern life and poor health habits? While most people think these mental and metabolic relationships are easy to explain, when you get down to the specifics of human physiology, things get significantly murkier.

When providers talk to people about changing their health behaviors—eating less or exercising more—they often get similar answers: "It's too hard," or, "I don't have enough energy." These answers are almost always met with strong disapproval. They are viewed as excuses for laziness, or signs of not taking the issue seriously enough, or of a lack of discipline. But is it possible that instead of being excuses, answers like "It's too hard" and "I don't have enough energy" are actually clues giving us important information? Could inertia and lack of motivation be *symptoms* of a metabolic problem? Is it possible that these people *literally* don't have enough energy?

As it turns out, it's not just possible; there's an abundance of evidence that it's true. You see, metabolism involves the production of energy inside cells. As you'll see in the coming chapters, people who have metabolic or mental illnesses have been found to have deficits in energy production inside their cells. These people are telling the truth. They really don't have enough energy.

It's not a motivational problem. It's a metabolic one.

We have been missing the elephant in the room.

Let's do a quick recap.

- I've described the current state of affairs in the mental health field and why what we're doing isn't working.
- I've explored evidence about the overlap and commonalities among mental disorders, as well as the limits of our current methods of differentiating between diagnoses. We've seen that every

mental disorder leads to a much greater probability of developing another mental disorder—any one of them. These bidirectional relationships suggest that *one common pathway* might be involved in all mental disorders.

- I have also explored the evidence of bidirectional relationships between mental disorders and at least three metabolic and at least two neurological disorders: obesity, diabetes, cardiovascular disease, Alzheimer's disease, and epilepsy. This raises the possibility of one common pathway not just for mental disorders, but for *all* of these disorders.

Already, this may seem impossible to reconcile. You might be yelling, "*But these are all different diseases!*" Schizophrenia is not the same thing as an eating disorder or a mild anxiety disorder. Cardiovascular disease, bipolar disorder, epilepsy, diabetes, and depression are all different. They have different symptoms. They affect different parts of the body. They appear at different ages. Some of them, like a stroke, can kill people quickly. Others, like mild depression that lasts only a few months, might come and go away without any intervention.

It's difficult to imagine that all these disorders share one common pathway. If there *is* such a common pathway, it will have to be involved in many different aspects of how the body works. It will need to tie together everything that we already know about these different disorders—their risk factors and symptoms, the treatments that work. That's a huge role for any bodily process or function to fill.

As you will see in Part Two, metabolism fills this role.

Yes: we have arrived at our common thread, the underpinning factor that lets us answer our tangled questions about causes and treatments, symptoms and overlaps.

Mental disorders—all of them—are metabolic disorders of the brain.

Part II

Brain Energy

Chapter 5

Mental Disorders Are Metabolic Disorders

The following 1938 observation from physicists Albert Einstein and Leopold Infeld is critically important:

> *Creating a new theory is not like destroying an old barn and erecting a skyscraper in its place. It is rather like climbing a mountain, gaining new and wider views, discovering unexpected connections between our starting point and its rich environment. But the point from which we started out still exists and can be seen, although it appears smaller and forms a tiny part of our broad view gained by the mastery of the obstacles on our adventurous way up.*[1]

For any new theory to be taken seriously, it must incorporate what we already know to be true. It can't just replace it; it must tie together our existing knowledge and experience into a broader understanding—one that will widen our perspective and offer new insights.

Mental health professionals have various camps at the base of Einstein and Infeld's mountain. Some believe mental illnesses are biologically based, that they arise from a chemical imbalance. They prescribe medications and have seen them work. Other professionals are focused on psychological and social issues. They have helped people through psychotherapy and social interventions and have seen these treatments work. They know with certainty that at least some mental disorders involve psychological and social issues; correcting these issues without any pills can solve the problem, at least for some patients. In reality, all these perspectives are correct. That this is so—and how it is so—can be seen clearly from the vantage point of our new theory: the theory of brain energy. This theory is based on one overarching concept—*mental disorders are metabolic disorders of the brain.*

In the medical world, new theories allow us to better understand the connections between treatments and disorders that we currently can't explain. They help us better predict future research findings. And they help us develop more effective treatments for the future. The theory of brain energy will do all of this for mental disorders. But its impact is bigger than just mental health. This theory ties together medical disciplines that most people think are unrelated—psychiatry, neurology, cardiology, and endocrinology. Others, too. All these disciplines have their own camps at the base of the mountain as well. Sometimes they work with each other and the practitioners see the connections between disciplines, but far too often, they don't. A patient might visit a cardiologist who prescribes their heart medicine, an endocrinologist who manages their diabetes prescriptions, and a psychiatrist who prescribes medication for bipolar disorder, with these specialists never communicating with each other. I'm hopeful that the brain energy theory will change this, leading to better cross-specialty collaboration and more effective and comprehensive care. Given what we've already seen about the connections between these disorders, this kind of communication and collaboration seems only logical. It may soon be possible to treat or prevent *all* of these conditions with one integrated treatment plan.

In order to prove, or at least strongly support, the theory of brain energy, the upcoming chapters will show that:

- Metabolic abnormalities have consistently been found in people with mental disorders, even those who don't yet have the already-recognized metabolic disorders of obesity, diabetes, or cardiovascular disease.
- Essentially all the risk factors for mental and metabolic disorders are the same. The list includes biological, psychological, and social factors, ranging from things like diet and exercise, smoking, drug and alcohol use, and sleep . . . to hormones, inflammation, genetics, epigenetics, and the gut microbiome. The list also extends to relationships, love, having meaning and purpose in life, and stress levels. You can isolate any one of these factors and see that it's been found to increase risk for both metabolic and mental disorders.
- Every one of these risk factors can be tied directly to metabolism.
- All the symptoms of mental disorders can be tied directly to metabolism, or more specifically, mitochondria, which are the master regulators of metabolism.
- All current treatments in the mental health field, including biological, psychological, and social interventions, likely work by affecting metabolism.

As we explore these lines of evidence, not only will it become clear that mental disorders are indeed metabolic disorders of the brain, it will become clear *why* this is important and what it means for treatment.

The Metabolic Ripple Effect

To suggest that so many different disorders stem from metabolic problems may sound far-fetched. Interestingly enough, while the medical field now groups obesity, diabetes, and cardiovascular disease together as metabolic disorders, that was not always the case. After all, they have very different symptoms, and they require different medications and different treatments. There are still different specialties that focus on these different disorders—obesity medicine (obesity), endocrinology (diabetes), cardiology (heart attacks), and neurology (strokes). However, they all affect the entire body, and people who have one such disorder are at higher risk of having another one. Not everyone who is obese has a heart attack or diabetes. Not all diabetics are obese. Not all people who have a stroke have diabetes. But while different people have different signs and symptoms, they are all interconnected.

The effects of metabolic disorders on the body aren't limited to an increased risk of the other metabolic disorders like obesity, diabetes, heart attacks, and strokes. As we've already discussed, these people have increased rates of Alzheimer's disease, epilepsy, and mental problems, too. But people with metabolic disorders are also more likely to develop innumerable *other* illnesses not usually viewed as metabolic. These include liver problems, kidney problems, nerve problems, brain problems, hormonal problems, joint problems, gastrointestinal problems, autoimmune problems, and even cancer.

Most people think that metabolic disorders are simple problems with simple solutions. They think they know the "root causes" of these disorders—behaviors like eating too much, not exercising enough, and/or smoking cigarettes. As long as people don't overeat, or under-exercise, or smoke, they will be perfectly fine, metabolically speaking. See? Simple.

But when it comes to metabolism, nothing is simple.

Let's look at an example. Mark was a seemingly healthy, thin, and fit forty-five-year-old man who developed multiple sclerosis (MS), an autoimmune disorder. To treat his MS, he was prescribed a medication called prednisone, a corticosteroid. Within weeks, he was bloated and gaining weight. Within a month, he developed pre-diabetes and was prescribed a diabetes medication. Unfortunately, weight gain and high blood sugar are both known side effects of prednisone.

Over the next six months, Mark gained forty pounds. Not all of this weight came out of nowhere; his behavior—specifically, his diet and exercise routine—changed dramatically. Up until his diagnosis, he had always eaten well and exercised vigorously several times a week. But corticosteroids like prednisone are known to increase appetite, and Mark had started craving and eating a lot of junk food, something he'd never done before. He tried to keep up with his exercise routine, but as he gained weight, it became harder and harder. He still managed to exercise some, but it was nothing like before. Mark's risk markers for cardiovascular disease worsened, including an increase in his blood pressure and lipids. He was now well on his way to a heart attack or stroke. Oh . . . and he also developed anxiety and mild depression. But who wouldn't be depressed and anxious in his situation? His doctor told him to try yoga and go on a diet. Unfortunately, that advice wasn't all that helpful.

What is the root cause of Mark's metabolic disorders? Within six months of starting the corticosteroid, he developed diabetes and obesity. The evidence is overwhelmingly clear that the culprit was the medication, not Mark's willpower or discipline. His cravings and lack of energy are symptoms of metabolic dysfunction. His depression and anxiety are also known side effects of this medication. In some ways, he's lucky that he didn't get manic or psychotic, which are also possible side effects.

Reactions like Mark's happen all the time with medications like prednisone. Other medications can cause these kinds of metabolic problems, too, including many psychiatric medications. But the takeaway here isn't that one should never take these medications: autoimmune disorders like Mark's can cause permanent organ damage, and the side effects of treatment are often determined to be a price worth paying when weighed against the severity of a disease. The point is that metabolic problems are not simple, nor are they avoidable through sheer willpower. Medication is only one of many, many possible causes. For example, a person who has experienced horrible childhood abuse is likely to have altered levels of cortisol, the body's equivalent hormone to prednisone. Perhaps unsurprisingly, people with a trauma history are more likely to develop metabolic disorders . . . and mental disorders, too. And once people develop a metabolic problem, symptoms and lifestyle changes like Mark's aren't unusual at all.

What Is Metabolism?

When most people hear the word "metabolism," they picture our bodies burning fat and calories. The common wisdom is that people with a "high metabolism" are skinny and have trouble gaining weight, while people with a "low metabolism" are overweight and gain weight easily—even if they don't eat that much. That's where metabolism begins and ends for most people.

Metabolism is about so much more than burning calories, though that's part of it. It influences every aspect of the way our bodies function.

To make energy, our bodies need food, water, vitamins and minerals, and oxygen—we breathe in oxygen and breathe out carbon dioxide, a waste product of metabolism. When we eat food, it gets broken down into carbohydrates, fats, and amino acids, along with the vitamins and minerals that are hopefully also there. All of this is absorbed into our bloodstream and shuttled around the body. Once nutrients arrive at cells and enter them, they are used as building blocks for things like proteins or membranes. Some might be stored as fat for a rainy day. But most of these nutrients will be converted into adenosine triphosphate (ATP), which is the primary energy molecule of the cell. ATP makes the machinery of the cell work.

That's the basic, high school biology version of metabolism. A one-sentence definition might be the following: *Metabolism is the process of turning food into energy or building blocks for growing and maintaining cells, as well as the appropriate and efficient management of waste products.* Metabolism is how our cells work. Our metabolism determines our cellular health, how our bodies and brains develop and function, and how we allocate resources to different cells at different times to optimize our survival. Metabolism allows some cells to grow and thrive and lets others shrivel up and die in a complex cost-benefit analysis that prioritizes healthy and advantageous cells over those that might be old, weak, or simply more expendable. As the body's resource management system, metabolism is all about adaptation. Our environment is constantly changing, and so are our circumstances in the environment. As a result, our metabolism is constantly changing to keep up with the shifts around us. These adaptations in metabolism allow us to thrive in optimal environments or to simply survive in situations that are stressful to the body, like food scarcity. But the availability of food isn't the only change metabolism responds to—numerous other factors play a role, like psychological stress, light exposure, temperature, how much sleep we get, hormone levels, and the amount of oxygen available to cells. At the end of the day, metabolism is the body's battle to stay alive. Many biological authorities would say that metabolism defines life itself.

Energy Imbalances

Metabolism is how our body creates and uses energy. And we can think of problems with metabolism as *energy imbalances.*

Problems with metabolism lead to problems in the way cells function. This goes for all cells in the human body. For instance, when heart cells are metabolically compromised, they don't work as well to pump blood. Brain cells need precise control. They need to be turned on at appropriate times and then turned off at appropriate times. When brain cells are metabolically compromised, these on/off processes can be disrupted. Precision is everything when it comes to brain function, and as we will see, this disruption can result in what we know as symptoms of mental illness.

The brain is the most complicated organ in the human body. In fact, the adult human brain is estimated to have about one hundred billion neurons. On top of that, there are an additional ten to fifty glial cells per neuron. The neurons are "nerve cells," and the glial cells are often thought of as support cells to the neurons. Combined, there are about one trillion to five trillion cells in the human brain. One group of researchers challenged this estimate, suggesting it's closer to eighty-six billion neurons and eighty-four billion glial cells, or 170 billion cells total.[2] Regardless, it's a lot of cells!

What coordinates the function of all these cells? Many people would say neurotransmitters, the cell's messenger chemicals. We can think of neurotransmitters as either "go" signals or "stop" signals, often categorized as excitatory (go) or inhibitory (stop). There are other variations, but these distinctions will suffice for now. They have been the primary focus of neuroscientists and biological psychiatrists for decades. But what controls neurotransmitters? How do cells know when to release them? Many would say their release is triggered by neurotransmitters from other cells. I'm sure you see the problems with this answer. It is partially correct. However, as I will discuss throughout the remainder of this book, there are numerous other factors that dictate the action of brain cells.

We've established that cells need energy to work. This energy is used for all sorts of different things throughout the body, including making muscles function, creating and regulating hormones, and making and releasing neurotransmitters. The parts of the body that need the most energy tend to be the parts of the body most affected by metabolic problems. As you might imagine, topping the list are the brain and the heart.

Although the brain makes up about 2 percent of the body mass, it uses about 20 percent of the body's total energy at rest. Brain cells are exquisitely sensitive to disruptions in energy supply, and when there's a metabolic problem somewhere in the body, the brain usually knows. Given that our brains are the control centers of our bodies, they ultimately control our perceptions of reality. When there is a metabolic problem somewhere in the body, we might perceive pain, experience shortness of breath, or feel fatigued or lightheaded. If there is a metabolic problem in the brain itself, the signs and symptoms can take just about any form. Sometimes they are obvious, such as confusion, hallucinations, or a complete loss of consciousness. Other times, they are more subtle, like fatigue, trouble concentrating, or mild depression.

Sometimes metabolic problems are *acute,* meaning that they are abrupt and dramatic. These can take the form of a heart attack, a stroke, or even death. A heart attack, for example, is usually due to a blood clot in one of the arteries feeding the heart. Some of the heart cells stop getting enough blood and oxygen. This prevents them from producing enough energy. If blood flow isn't restored quickly, the heart cells die. This is a metabolic crisis in the heart. A stroke is an acute metabolic crisis in the brain. The ultimate metabolic crisis is death itself, where the cells in the entire body stop producing energy. Many paths can lead to this whole-body energy failure—heart attacks, strokes, poisonings, severe accidents, cancer. They all result in the cells of the body no longer producing enough energy, and it is the lack of energy production that results in death.

Heart attacks, strokes, and death are all examples of absolute and acute energy problems that result in cell death. However, there are less dramatic situations in which the energy supply to cells is compromised: Instead of a complete shutdown of energy production, the cells simply aren't getting enough energy. Instead of cell death, the cells don't work quite right. Some of these metabolic problems can last for just a few minutes, while others might last for hours. Hypoglycemia, or low blood sugar, is a good example. It most commonly occurs when people haven't eaten in a while. In mild cases, it results in feeling hungry, irritable, fatigued, or having trouble concentrating. In moderate cases, it might result in a headache or feeling depressed. In severe cases, it can result in hallucinations, seizures, or coma. If it progresses further, it *can* result in absolute metabolic failure—death. Before things get so severe, however, most people implement the obvious solution—they eat something. This raises blood sugar, and things begin to function normally again. Even if they don't eat anything, the body has systems in place that usually prevent severe hypoglycemia. However, for diabetics who inject insulin or take medication to more forcibly lower blood sugar, these severe consequences are a real possibility. You might notice that brain symptoms dominate the above list of effects, even though the hypoglycemia is occurring throughout the entire body.

Other metabolic problems are not acute but instead *chronic* disorders with long-lasting symptoms—like diabetes, for example. Many people think of diabetes as high blood sugar. However, a paradoxical and interesting way to think about diabetes is as an energy shortage, or a deficit in energy production. Glucose is the primary fuel source for cells. In diabetes, cells have trouble converting glucose into energy. The levels of glucose in the blood can be high, sometimes very high, yet that glucose has trouble getting into cells where it can be used. Getting glucose from the bloodstream into cells requires insulin, a hormone produced by the pancreas. Diabetics have either a shortage of insulin or insulin resistance—a condition in which the body is not as responsive to insulin. When cells don't have enough glucose, they aren't able to produce enough energy. When they don't have enough energy, they don't work right.

Since glucose is the primary fuel source for most cells in the body, diabetes can affect many different parts of the body. But not everyone has the same problems. The symptoms of diabetes can be wide-ranging and can change over time. In the beginning, symptoms are usually mild. They can include things like urinating too much or losing weight unexpectedly. They can also include mental symptoms such as fatigue or trouble concentrating. As the illness progresses, different organs can be affected. Some people develop problems with their eyes, nerves, or brains. Some people have heart attacks or strokes. Others experience kidney failure or contract serious infections that are difficult to treat.

Why are the effects on people so different? Why don't all diabetics end up with the same symptoms and the same body parts failing? The answer is complicated—and often related to metabolism.

Metabolism is affected by numerous factors. It is always changing. And it is different in different cells of the body at different times. Some cells can be functioning normally while others are dying. Some cells may gradually malfunction as the result of chronic energy deprivation. Metabolism is not all-or-nothing. It is controlled at a variety of levels. Some of the factors that affect metabolism do so broadly, while others are specific to distinct parts of the body. Some are specific to specific organs. Some are specific to specific cells.

Metabolism Is Like the Flow of Traffic

Think of it like this—the body is like a large city with lots of roads and highways. There is a lot of traffic. Each car is like a human cell. During rush hour, it can be hectic. If you're in a car, it can feel chaotic. So many things to pay attention to: stoplights, cars changing lanes, someone on their cell phone swerving into your lane. If you look at the traffic from above, however—say, from the top of a skyscraper—it looks pretty orderly. The roads are organized. The cars and trucks are moving along. Some cars stop while other cars go. They wait their turn and then start going again. Cars go slow on certain roads but then speed up on highways. Some cars change lanes, and the cars around them have to slow down to let them in. Others might be having problems and be stuck on the side of the road. There might be some traffic accidents, causing other cars to take a detour. If you were to try to take in the specifics of each and every car at the same time, it would be overwhelming—there are too many cars, too many stoplights, too many other factors to keep in mind. But when you look at the big picture, traffic is moving along. The city is working. People are getting to where they need to go. The city is alive. It has energy; you can see it flow. This is the way to think about metabolism in the human body.

Back to the question I posed earlier—why do some people with diabetes have different symptoms? More important to the theory of brain energy, if all mental disorders are metabolic disorders, why doesn't everyone with a mental disorder have the same symptoms?

Illnesses and symptoms are like traffic jams. Either traffic isn't flowing optimally, or it stops altogether. One highway might represent the pancreas. An access road might represent a specific brain region that controls attention and focus.

What causes a traffic jam on a given road or highway? Myriad things. Car accidents, road construction, potholes, or traffic signals failing to work. The design and maintenance of the roads play a role, and the cars and drivers do, too. Some parts of the city have more frequent traffic problems. This can be due to poor design, poor maintenance, or more aggressive or careless drivers on those roads. The areas of the city with regular traffic problems represent "symptoms" or "illnesses"—places where the traffic isn't "working" properly.

When it comes to human illnesses and symptoms, we are talking about parts of the body or brain that aren't working properly. This is usually the result of an issue in one of three areas: the development, function, or maintenance of human cells. Cells must *develop* properly in order to meet the needs of the body. *Function* is making sure all the parts are doing what they are supposed to be doing, in the right ways at the right times. *Maintenance* is keeping everything in good shape. This is analogous to traffic needing adequate design and construction of roads and bridges (development); all the cars, drivers, and traffic lights working properly (function); and the whole system being regularly serviced—cars tuned up, roads patched, stoplights tested, etc. (maintenance).

In humans, these three things—development, function, and maintenance of cells—ultimately depend upon one thing: metabolism. If there are problems with metabolism, there will be problems in one or more of these areas. If the problems are significant enough, there will be "symptoms."

So, what affects metabolism? Just like traffic in the city, many things! Diet, light, sleep, exercise, drugs and alcohol, genes, hormones, stress, neurotransmitters, and inflammation, to name a few. However, each of these things affects different cells in different ways. Depending upon the mix of factors that someone is exposed to, different cells and organs will be affected, resulting in different symptoms and different illnesses. Just like some roads are more susceptible to traffic jams, some cells are more susceptible to metabolic failure. Sometimes, parts of the body will function normally at times of low demand but begin to malfunction at times of increased demand—just like a rush-hour meltdown on a city highway that's overwhelmed by commuters.

We've established that metabolism defines life itself; that it determines how cells function, that it affects and is affected by innumerable factors. In a way, of course mental disorders are related to metabolism. In essence, *everything* is! So what?

What I'll show in the coming chapters is that metabolism is, in fact, the *only* way to connect the dots of mental illness. It is the lowest common denominator for *all* mental disorders, *all* of the risk factors for mental disorders, and even *all of the treatments* that are currently used. And, perhaps most significantly, although metabolism is complex, solving metabolic problems is usually possible, oftentimes through straightforward interventions.

Before I dive into the evidence for all of this, however, I first need to clarify what a mental disorder *is* in the first place. This question has long plagued the mental health field, and it centers on one issue in particular—the difference between normal mental states (especially stressful and adverse ones) and a mental disorder.

Chapter 6

Mental States and Mental Disorders

As I discussed in Part One, one of the dilemmas in the mental health field is distinguishing between normal human emotions and mental disorders, especially since the symptoms can be the same. We all get anxious or mildly depressed from time to time. If we experience a devastating loss, such as the unexpected death of a spouse, we may get severely depressed for a period. These are all normal reactions. They are hardwired into our brains.

However, when people are exposed to numerous stressors all at once, or when the stressors are extreme or overwhelming (such as being violently assaulted), these normal and understandable initial reactions can quickly lead to what we call "mental illness." The diagnoses are all over the map. Trauma or extreme stress can lead to anxiety disorders, depression, PTSD, eating disorders, substance use disorders, personality disorders, and even psychosis. How do stress and trauma lead to all these different disorders? And where is the line between a normal reaction to adversity and a disorder?

Two issues have made these questions particularly difficult to answer: (1) the symptoms are the same, and (2) both mental states and mental disorders can lead to poor health outcomes. Nonetheless, distinguishing between normal mental states and mental disorders is critically important. Mental states are adaptive reactions to adversity. Mental disorders represent the brain malfunctioning. These distinctions have direct implications for treatment. Helping people cope with adversity is very different than treating a brain that is malfunctioning.

Understanding "Normal": Stress and the Stress Response

Stressors are the psychological and social factors in the biopsychosocial model—the ones that people usually think of as the "mental" causes of mental illness.

Many clinicians and researchers still see biological factors as separate from psychological and social ones. For example, they might believe that hallucinations are due to a biological chemical imbalance, but that someone with schizophrenia can also suffer from low self-esteem, which is a psychological problem. They may try to address both, but they often see these issues as unrelated and separate. One requires medication, and the other requires talk therapy. I disagree with this dichotomous view. I think biological, psychological, and social factors are all interconnected and inseparable. Biology influences our psychology and how we get along with others. But our psychology and our interactions with other people influence our biology. These connections can impact all mental and metabolic symptoms. To begin unpacking this, let me start by making some overarching observations about our species.

Humans are meant to live in groups. We seek out and attach to other people—parents, lovers, children, friends, teachers, and community members. These connections form a network of safety and support in our lives. We are biologically driven to want, and even need, these people. There is a conundrum, though: While we must live with other people, other people are actually the primary sources of psychological and social stress. Most of these stressors revolve around relationships, roles, resources, and responsibilities. People can be stressed over expectations on them, financial problems, performance problems, relationship problems, or status in society. Some people experience chronic stressors due to socioeconomic status, abuse, neglect, race, ethnicity, religious beliefs, physical abilities, cognitive abilities, gender identity, sexual orientation, age, and so many other factors. People can be harmed or threatened by other people. We sometimes make each other feel unsafe. We sometimes make other humans feel like they aren't good enough. There are countless reasons that humans stress other humans. And interestingly, the absence of other humans, or loneliness, is also a powerful stressor unto itself.

All of these stressors lead to the *stress response,* a complex array of biological changes in the brain and body. The stress response includes changes in four domains:

1. The hypothalamic-pituitary-adrenal (HPA) axis, which results in cortisol flowing through the bloodstream;
2. The sympathetic-adrenal-medullary (SAM) axis, which results in adrenaline (epinephrine and norepinephrine) flowing through the bloodstream;
3. Inflammation;
4. Changes in gene expression, especially in the hippocampus.[1]

All of these changes, in turn, *affect metabolism.* They comprise a person's response to adversity. They are not disorders. They set the stage for "fight or flight." However, in most day-to-day stressful situations, we don't fight or flee. Instead, we just stay put—but we get angry, or anxious, or irritable, or overwhelmed, or confused, or terrified, or hurt, or sad. Yet these core changes are still occurring in our bodies and brains.

Different stressful situations result in different behaviors and emotions. Some stressors make you want to yell at someone, such as the driver who cuts you off in traffic and then flips you off for no reason. Other stressors might make you ruminate and not sleep well, such as feeling unprepared for an important exam the next day. Others might make you want to curl up into a ball and cry, such as getting dumped by the love of your life. All of these situations involve the stress response. While similar mechanisms are involved, clear differences trigger different brain regions to create different responses.

Although these are normal, the stress response takes a toll—*a metabolic toll.* The body uses energy to produce these changes, meaning less energy is available for other functions. Many of these responses create a state of high alert. In some situations, the person feels threatened and prepares to fight or argue with someone. In other situations, the person may feel wounded, vulnerable, or powerless and try to hide from the world. In either case, metabolic resources are being mobilized. The heart is pumping faster. Blood pressure is increasing. Blood glucose is rising. Hormones are flowing. Inflammatory cytokines are being released. The body is mounting resources and energy for its own defense.

When the stress is mild, people who are resilient and metabolically healthy deal with it. It can be over in a matter of seconds or minutes.

However, if a body is metabolically compromised, or if the stress is extreme, people can be pushed over the edge, and a new mental or metabolic disorder can quickly emerge. For those with preexisting disorders, symptoms can become even worse. That's right: *Stress can exacerbate every known mental and metabolic disorder*. People with depression may get more depressed. People with alcoholism may fall off the wagon. People with schizophrenia may hallucinate. People with Alzheimer's disease may get agitated and combative. People with epilepsy may experience a seizure. People with diabetes may have their blood sugar skyrocket. And people with cardiovascular disease may have chest pain or a heart attack. Some people die—from stress alone. This is all well established.

A separate medical field has tried to make sense of all of this—the field of psychosomatic medicine or *mind-body medicine*. Many healthcare professionals have observed the relationship between psychological and social factors affecting the health of the body. Practitioners in this field understand that all these risk factors play a role in human physiology. These factors are often referred to as the *social determinants of health*. Many social factors, such as poverty, abuse, or living in a high-crime neighborhood, can have huge consequences on health and longevity.

Some of the most compelling data on this comes from the adverse childhood experiences (ACEs) studies that started between 1995 and 1997 and looked at the number of adverse experiences that children and adolescents encounter and their effects on long-term health outcomes, both physical and mental. These ongoing studies have looked at stressors early in life, such as physical and sexual abuse, neglect, household substance abuse, household mental illness, exposure to domestic violence, and parental divorce, and then determined if these early experiences were associated with later health outcomes. A 2017 meta-analysis of thirty-seven such studies looking at twenty-three health outcomes in more than 250,000 people found that they are.[2] The more ACEs a child has, the more likely he or she will have poor health outcomes. ACEs increase the probability for physical inactivity, obesity, and diabetes by 25 to 52 percent. They are associated with two to three times higher rates of smoking, poor self-rated health, cancer, heart disease, and respiratory disease. ACEs lead to a three- to sixfold increase in the rates of sexual risk-taking, poor mental health, problematic alcohol use, and illicit drug use. They also lead to more than a sevenfold increase in being the victim or perpetrator of violence, a tenfold increase in problematic drug use, and a thirtyfold increase in suicide attempts. ACEs clearly affect mortality. One study of 17,000 people that looked specifically at mortality data estimated that having six or more ACEs takes twenty years off a person's life compared to those with no ACEs.[3]

These studies have led many people to conclude that ACEs *cause* both physical and mental illnesses. Some experts have gone so far as to suggest that ACEs, in particular childhood trauma and abuse, are likely *the common pathway* to all mental disorders. But let me remind you, these are correlations. They don't prove causation. More importantly, not everyone who has a horrible childhood develops a mental disorder, and many people who end up with mental disorders have perfectly fine childhoods. Nonetheless, if these adverse experiences are playing some role in these different disorders, how does that work? What is happening in the body and brain to cause all of this?

Limited Resources

For decades, researchers have been studying the biological effects of stress on the brain and body to better understand these relationships, hoping to identify the cause-and-effect pathways from stressful life events to poor health outcomes.

We know that when the body is stressed, metabolic resources are being diverted to the fight-or-flight system. This leaves less energy available for other functions. Any cells that were already struggling can begin to fail. This can lead to metabolic and mental symptoms.

Stress also impairs the body's ability to maintain itself. Cells engage in housekeeping functions daily. They get rid of damaged cell parts, various waste molecules, and misfolded proteins, and they make new ones to take their places in a process often called *autophagy*. *Auto* means "self" and *phagy* means "eat," so this term literally means "to eat yourself." Our cells degrade these old parts in waste-disposal systems called lysosomes. Some of this material is recycled and used to make new parts. High levels of cortisol have been found to inhibit autophagy, slowing down or stopping this maintenance process.[4] Problems with autophagy have been found in a wide variety of disorders including neurodegenerative, neurodevelopmental, autoimmune, inflammatory, cancer, schizophrenia, bipolar disorder, autism, alcoholism, and major depression.[5] Disturbances in autophagy are known to affect neuroplasticity and the maintenance of brain cells.[6]

In addition to problems with autophagy, when cells are stressed, they also slow the process of making new proteins. This appears to conserve metabolic resources for the body's defense system. One way they delay making these proteins is by sequestering messenger RNA molecules (the instructions for new proteins) into little bubbles called "stress granules."[7] These have been associated with neurodegenerative disorders, and high levels of cortisol stimulate their production.[8]

One additional way that stress can lead to maintenance problems is through sleep disruption. It is well established that stress can lead to insomnia. Sleep is critically important to both physical and mental health. It's a time when the body prioritizes maintenance functions. When people aren't sleeping well, their bodies aren't doing this maintenance work. On top of that, sleep deprivation, in and of itself, is stressful and can lead to higher levels of cortisol, which can make the problem even worse.

All this stress, regardless of when it occurs, results in premature aging. I've already mentioned that all *mental disorders* are associated with premature aging, but *stress alone* can also cause it. One study tried to quantify the effects of stress on aging.[9] The study recruited fifty-eight *healthy* premenopausal women who were mothers to either healthy or chronically ill children. The women's average age was thirty-eight, and they did not yet have identified health problems. The researchers assessed three metrics of aging and asked the mothers to rate their perceived levels of stress. The mothers with the highest levels of stress over the longest periods of time showed signs of accelerated aging compared to the lowest-stress women. On average, they aged ten years faster.

Stress clearly plays a role in human health and takes a serious metabolic toll. It uses energy that could otherwise be used for proper cell function and maintenance. When people are stressed in extreme ways or for prolonged periods of time, their bodies can get worn down and begin to malfunction, resulting in various physical and mental disorders, or just plain old aging. If the brain or body are already compromised and vulnerable, stress can make symptoms worse, because the energy needed for the stress response is diverting energy from these vulnerable cells.

Stress-reduction practices, such as mindfulness, meditation, or yoga, can play a powerful role in treatment (more on that in Part Three). However, they aren't the solution for everyone. If a person is living in an adverse environment, turning off the stress response may not be possible or even advisable. Soldiers fighting in a war are in danger. And while their service directly results in greater risks for mental and metabolic disorders, their heightened stress response is protecting them. The same applies to people in dangerous neighborhoods. Instructing people in dangerous environments to take deep breaths and be mindful isn't the full answer. When they get to safety, these strategies might play a role, but the damage may already be done by then.

Furthermore, stress may not be the cause of one's mental illness at all. In that case, stress-reduction techniques likely won't be all that helpful.

Understanding "Disorders": A New Definition of Mental Illness

As I discussed in earlier chapters, the current classification of mental disorders is fraught with problems—heterogeneity, comorbidity, and a lack of validity. None of the diagnoses are true and distinct disorders unto themselves.

The NIH has recognized this for some time and developed a new framework for thinking about mental illnesses—the Research Domain Criteria (RDoC). The RDoC starts fresh, ignoring our current diagnostic labels and classifications. Instead, the framework focuses on domains of functioning—emotion, cognition, motivation, and social behavior. It assumes ranges in these constructs from normal to abnormal, and it encourages researchers to explore these constructs from a perspective other than a diagnostic label. At one point, proponents of the RDoC were calling for a complete overhaul of our current psychiatric diagnostic criteria. Changing psychiatry and the mental health field, however, is no easy feat, so our current diagnostic criteria remain, despite all the known flaws. RDoC remains only in the realm of research at this point. However, for our purposes, I am going to use this model to define mental illness in the context of the brain energy theory.

It starts by setting aside DSM-5 diagnostic labels and focusing instead on symptoms. This doesn't mean that some of the diagnoses aren't useful. Many are. Our current diagnostic labels simply describe some of the more common ways that the brain malfunctions. After all, the brain works, or fails to work, in predictable ways, and we can use those common narratives to our benefit.

The human brain is like a machine—a very sophisticated and complicated machine, but a machine nonetheless. It has many parts all designed to do certain things. Some are fairly straightforward, such as making our muscles move or sensing what we feel or see. Other functions of the brain are more complicated, like sophisticated computer algorithms that are triggered in certain situations. In one way or another, all these brain functions can be tied to helping us survive, adapt to our environments, or reproduce.

Given that the human brain has billions, if not trillions, of cells, and that each cell is a complicated machine itself, we face a potentially overwhelming problem: with so many cells, it seems there's an almost infinite number of ways that all of these "parts" could malfunction. For better or worse, this is where the mental health field has been focused, with researchers trying to understand how the machine works, step-by-step. It's an overwhelming task, the notion of wholly mapping something as complicated as the human brain, and waiting for this work to be finished has arguably limited our progress in better understanding and treating mental illness.

But it doesn't have to be so complicated. It turns out that all symptoms of mental illness actually correspond to normal mental states or brain functions, but gone awry: present when they should not be, absent when they should be present, or more or less active or persistent than is appropriate. These brain functions include things that relate to emotions, cognition, behavior, and motivation. As I will discuss, even some of the more bizarre-seeming symptoms of mental illness, such as delusions and hallucinations, can be tied to normal brain functions. Although we don't know precisely how all of these functions work, we know they exist. That's enough for our purposes here.

Let's start with a simple definition, then: Mental illness is when the brain is not working properly. Normal brain functions are either *overactive, underactive, or absent.* An easy example is having a panic attack for no clear reason. The panic system is beneficial when facing danger. It gets us moving. When it gets triggered for no clear reason, it's dysfunctional and maladaptive. Sometimes the opposite can occur—brain functions that fail to activate in the right situations. Consider memory impairment in someone with dementia or a lack of social skills in someone with autism.

When it comes to symptoms of mental illness, many people would say that they can't possibly correspond to normal brain functions. It can seem like somehow the brain is doing unique and highly unusual things for no clear reason. I look at it differently. Like the parts of any machine, the parts of the brain are either working or they aren't. If they perform their usual function, but get turned on at the wrong time, it can result in symptoms that seem bizarre. The same is true when normal brain functions fail to activate, or if two unrelated brain functions are mistakenly occurring at the same time.

A Simple Example: Three Cars

Let me use an analogy to explain how I think about the way people with mental illness differ from those who are having "normal" stress reactions, even though their symptoms can be the same and they can both lead to poor health outcomes. I'll describe three cars. Each is the same make and model, so in theory, they should have the same lifespan and overall "health." Each represents a human being.

Car A lives in California, where the skies are blue, and the roads are in great condition. The owner doesn't drive much—maybe twice a week. Car A is housed in a garage and gets regular maintenance. Car A is living the good life!

Car B lives in the mountains of New Hampshire, where the winters can be fierce, and the back roads are filled with potholes. The owner drives the car every day and doesn't have a garage to store it in. When winter comes,

Car B gets snow tires, and sometimes even snow chains. In a blizzard, Car B is using its headlights, windshield wipers, blinkers, snow tires and chains, and its four-wheel drive system. The brakes are applied often so that the driver doesn't lose control. In these situations, Car B is getting very low gas mileage compared to Car A. Car B also has more maintenance problems, given the harsh winter environment and difficult driving conditions. In the end, Car B has more "health problems" and ends up living a shorter life than Car A.

Cars A and B are "normal"—they are both doing what they are supposed to, given the environmental circumstances they're in. Neither has a disorder. Car B has more health problems and ends up living a shorter life, but given the adversity it faces, this is normal. The adaptations it uses, such as snow tires and chains, four-wheel drive, and frequent braking, are like stress responses—depression, anxiety, fear, anger. They help Car B navigate its difficult environment and they serve extraordinarily useful purposes. Without them, Car B would be much worse off.

Now let me tell you about the third car. Car C lives in Indiana, where the weather is less harsh than New Hampshire, and the roads are in decent shape. It gets driven five days a week, sometimes in good weather and sometimes in bad. But Car C has problems. It turns on its windshield wipers and blinkers even in sunny weather. The wiper blades are worn thin because they get used so much. They end up scratching the windshield. Car C sometimes uses its four-wheel drive and travels only twenty-five miles per hour on the highway, even though it's sunny and all the other cars are going sixty. When Car C drives at night, it doesn't turn on its lights, even though they are needed. Car C has a disorder comparable to a mental illness. Although it has the exact same features and adaptive strategies as Cars A and B, it is using some of them at the wrong times and under the wrong circumstances. Meanwhile, it's failing to use others that it should be using. Car C ends up needing quite a bit of maintenance. It also gets into traffic accidents. Car C's disorder is seriously affecting its health and safety and impacting its ability to get along with other cars on the road. Car C ends up dying an early death.

So . . . Cars A and B are "normal," and Car C has a disorder.

People who are struggling with adversity, like Car B, often need help, even if their brains are not malfunctioning. Their biology is responding to their adverse life experiences in "normal," predictable, and adaptive ways. To aid them, we need to change their environments or help them respond optimally to harsh conditions. For the most part, these are societal factors—things like war, poverty, food insecurity, abuse, systemic racism, homophobia, misogyny, sexual harassment, anti-Semitism, and many other societal "blizzards." Changing society so that these blizzards no longer exist is the ideal way to deal with these issues. In the meantime, helping people cope as best as possible can also be helpful.

The brains of people with mental disorders are malfunctioning. They are doing things at the wrong times or with the wrong intensity, or they are failing to do things that they should be doing—as in the case of Car C. You don't need to know precisely how these work in order to determine if there is a problem or not, just like you don't need to fully understand the inner workings of a car and its windshield wiper system to know if there is a problem or not. I imagine you're thinking that the problem with Car C isn't with the car itself, but with the driver of Car C. In fact, you're correct. I'll get to that soon enough.

It's important to point out that prolonged or extreme stress *can* lead to a disorder as well. At some point, Car B could easily develop maintenance problems that result in adaptive strategies no longer working—maybe the lights stop working or the windshield wipers get worn thin and are no longer effective (underactive functions). Or the blinkers won't turn off (overactive function). At that point, Car B would also have a disorder.

A Human Example: Pain

Now I'll try to show you that this really does happen in the human body by focusing on an easy, straightforward example—pain. Since pain is controlled by nerve cells and brain regions, it serves as a perfect example for most of the mental symptoms I'll discuss.

Pain is a normal, healthy experience for humans—even though it is quite unpleasant. It saves our lives. It protects us from injuring ourselves. Pain is controlled by pain receptors, a nerve that goes to the spinal cord, another nerve that goes up to the brain, and then the regions of the brain that sense and process pain. The function and dysfunction of these neurons and brain regions gives us a simple framework that will help us better understand mental disorders.

Broadly speaking, disorders of the pain system can be lumped into three categories based on the function of the cells in the pain system—overactive, underactive, and absent.

1. **Overactivity** of the pain system is when people experience pain more frequently or intensely than they should. Clinicians and researchers will often describe this as *hyperexcitability* of the pain system. For example, people with diabetes can develop neuropathy, and the nerve cells or brain regions that process pain can fire when they shouldn't or fail to turn off when they should. This causes pain even when nothing painful is happening. It can result in a chronic and debilitating pain condition for some people.

2. **Underactivity** can occur when people feel fewer pain signals than they should, which can also happen in people with diabetes. In addition to hyperexcitability, diabetic neuropathy can also result in reduced sensation, especially in the feet. The nerves aren't working properly, and this results in underactivity of the pain system. We know the nerves are still there and alive because sometimes people feel something.

3. **Absence** of pain can occur with prolonged and severe diabetes, but also with other conditions like spinal cord injuries or strokes. People feel absolutely nothing because the cells are dead or severely injured and no longer working.

These three scenarios—overactivity, underactivity, and absence of function—are all disorders. The pain system isn't working correctly.

In some cases, it can be difficult to draw a line between normal pain and a pain disorder. One example is a herniated disc in the lower back causing pain. When the disc first becomes herniated, it's not a disorder. The pain system is doing what it is supposed to do. If the pain goes on for a prolonged period of time, however, even after surgery and multiple medications, at some point we label it a pain disorder. What makes it a disorder? The nerves can become injured from the pressure of the herniated disc. These injured nerves can become hyperexcitable. They can send pain signals too often or too intensely. The point at which the pain goes from a normal response to a disorder is difficult, if not impossible, to distinguish based on current diagnostic tests. In some cases, it's not clear whether it's normal or a disorder. However, when the pain becomes chronic, severe, and unprovoked, we call it a disorder.

Regardless of whether pain is a normal response to an injury or a pain disorder, treating the pain is appropriate all the time. For example, we all know that people will feel pain when they get surgery. It's normal and expected. However, we still treat it to alleviate suffering.

This distinction between normal and abnormal is important. Doctors who treat pain need to have good clinical skills. They need to understand the many reasons that someone might be experiencing pain. They need to evaluate their patients for these causes before assuming that they have a pain disorder. If a patient comes in with foot pain, it might be due to a sprain, a muscle spasm, a broken bone, or a piece of glass wedged in the skin. Each cause requires very different treatments. Treating the pain as though it is due to pain disorder might bring some relief, but it won't solve the problem. In fact, the problem may get worse. However, if no obvious causes for foot pain are present, the doctor might then diagnose a pain disorder. This same type of detailed cause-and-effect assessment is necessary when evaluating people for mental disorders. Again, helping people cope with adversity is very different than treating a malfunctioning brain.

Back to Defining Mental Illness

Here's our new, simplified definition of mental illness: *A mental illness is when the brain is not working properly*. Now let's expand on that definition: *A mental illness is when the brain is not working properly over a period of time, and this causes mental symptoms, which lead to suffering or impairment in functioning.*

Although this is a fairly short and concise definition, every part matters, and no part can be taken out of context. This definition includes four necessary components:

1. The brain is not working properly.
2. This results in mental symptoms.
3. This malfunction occurs over a period of time.
4. The symptoms cause suffering or impairment in functioning.

Although these might seem like simple concepts, they can quickly get complicated.

The first component of this definition—*the brain is not working properly*—sounds straightforward. But it's actually difficult to measure and assess based on current technologies, just like pain. We have many tests that can measure brain health and function, such as EEGs and neuroimaging studies. However, none of them is sensitive and specific enough to accurately diagnose a mental disorder. Measuring the function of microscopic brain regions is difficult. So, in the real world, how do we know if the brain isn't working properly?

That leads us to the second component of the definition—*this results in mental symptoms*. Symptoms are the best indicator of abnormal brain function. However, like pain, when it comes to symptoms of mental illness, most can be normal, healthy brain functions in the right circumstances. Even something like hallucinations can occur in most people in the right circumstances. We all have hallucinations when we dream—we see things and hear things that aren't there. When these things occur at the wrong time or when they fail to activate at the right time, they might represent a disorder. We can categorize the symptoms into the same three basic categories I used for pain—overactivity, underactivity, and absence of function.

Component three—*this malfunction occurs over a period of time*—reinforces that the duration of symptoms matters. All of our brains fail to work perfectly at least some of the time, and this results in what we might call symptoms. Most of us have occasional lapses in our memory. Sometimes, we think we hear a noise, but no one else hears it. Sometimes, we "wake up on the wrong side of the bed" and feel depressed for no clear reason. These are examples of the brain not working properly. These are not mental illnesses but common occurrences that can happen due to a variety of circumstances—a bad night's sleep, an extraordinarily stressful situation, use of alcohol or drugs, or just having a bad day. These are usually short-lived experiences (also relating to metabolism) with an easy brain-body fix. Mental illnesses need to be *persistent* problems with brain function that result in symptoms. The persistence of symptoms is currently part of our diagnostic process in the mental health field, but the amount of time varies by diagnosis.

That brings us to component four of our definition: *the symptoms cause suffering or impairment in functioning.* We all have changes in emotions, cognition, motivation, and behaviors over the course of our lives. We learn. We grow. We meet new people and make changes. We go through challenging experiences. We suffer losses and setbacks. These fluctuations alone are not mental illness. Only when a person is distressed by these changes in an unusual way, or the changes prevent them from functioning in life, do we begin to consider the possibility of a mental illness. There is no question that this part of the definition is tricky, and the debates around the issue of suffering and impairment in functioning are complicated. Two issues are particularly important:

1. **People have the right to be unique, creative, make changes in their lives, and go against mainstream culture.** Being different is

not a mental illness. Yet other people's rejection of uniqueness may cause suffering. For example, many teenagers go through a rebellious phase. This is often a normal part of growing up and separating from parents. This alone is usually not a mental illness. Many people go on diets and frequently check their weight. They begin to think more about what to eat and how they look. This is not automatically an eating disorder. Both situations involve changes in emotions, cognition, motivation, and behaviors, but unusual distress and the inability to function aren't part of the picture.

2. **Some people with mental disorders lack insight.** They don't realize that their symptoms are somehow abnormal. They don't recognize how their symptoms affect their behavior and function. They have trouble understanding why other people perceive these changes as unusual. They may claim that they are perfectly normal. However, if their symptoms are seriously impairing their ability to function in society, then a mental illness needs to be considered.

   It is common for people with hallucinations and delusions to lack insight into their illness. For instance, people with paranoia will say that they are really being persecuted—it's not "mental," it's real. People with eating disorders will sometimes talk about how happy they are to be losing so much weight and looking better. They view any changes in functioning, such as devoting less time to school or friends, as the sacrifice needed to lose weight and look good. They may ignore the serious health problems that are evident to everyone else. Both will claim that their changes in emotions, cognition, motivation, and behaviors are normal and expected for anyone going through their circumstances. They will often deny any impairment in functioning. So, are these mental disorders? Yes. They are causing significant distress and/or impairment of functioning (this would include health problems), even if the person doesn't see it or acknowledge it.

These nuanced dilemmas sometimes make it difficult, if not impossible, to distinguish between being different and living in an unforgiving and rigid society versus having a mental illness. The mental health field has changed its stance around issues like this over the years, such as labeling homosexuality a disorder at one point in time and then reversing that decision.

Symptoms of Mental Illness

Now that we've arrived at and unpacked a new definition of mental illness, let me put it into action by outlining three broad scenarios that can produce symptoms of mental illness. They follow the model that I outlined for disorders of the pain system—brain functions that are overactive, underactive, or absent.

Overactive Brain Functions

Overactivity or hyperexcitability of brain cells and networks has been documented in many mental disorders. When thinking about this phenomenon, we are looking for symptoms or brain functions that are occurring more often or more intensely than they should be, or at the wrong time.

Fear and anxiety symptoms can result from hyperexcitability of the amygdala—one of the regions of the brain implicated in the fear response. These neurons may fire out of turn or not stop firing, which causes anxiety symptoms at inappropriate times or an exaggerated fear response.

Obsessions and compulsions can result from hyperexcitable cells and networks in the brain areas associated with grooming and checking behaviors. We all normally groom ourselves and check things. OCD occurs when these systems are overactive.

Psychotic symptoms, such as hallucinations and delusions, are found in many mental and neurological disorders. They also occur in many people who never get diagnosed with any disorder.

The precise brain cells and regions that cause psychotic symptoms are currently unknown, despite decades of intensive research looking for them.

Nonetheless, there are a few ways that we can think about what might be happening in the brain.

The easiest way to understand psychotic symptoms is hyperexcitability of brain cells that process perceptions. For example, if the brain cells and networks that perceive sound are hyperexcitable, people will hear something that isn't there—an auditory hallucination. Neurosurgeons can make people "hallucinate" by stimulating brain areas with an electrode. Hyperexcitable cells would be doing essentially the same thing.

The problem might not be in the neurons that perceive sound but in other neurons that regulate them and slow them down. There are a group of neurons called "cortical interneurons." These neurons are known to be inhibitory, as they secrete gamma-aminobutyric acid, or GABA, a neurotransmitter that slows activity in its target cells. Abnormalities in the function of these neurons have been found in many disorders, including schizophrenia, Alzheimer's disease, epilepsy, and autism. This lack of inhibition would result in overactivity of the neurons they are supposed to be inhibiting.

Another possibility is that psychotic symptoms are related to the sleep systems in the brain. As I mentioned, we all have hallucinations and delusions every day—in our sleep. When we dream, we hear things and see things that aren't there. We can believe wild and crazy things. Many people have nightmares that include being chased or persecuted. If these experiences occur during sleep, they are just bad dreams, not mental disorders. It's possible that the same brain cells and networks that create these experiences at night are hyperexcitable and firing erroneously during the day in people with mental disorders.

For some delusions that can seem bizarre, such as Capgras syndrome in which people believe that their loved ones have been replaced by imposters, we actually do know some of the specific brain networks involved in this process.[10] These areas of the brain appear to be overactive and/or underactive.

One important observation is that hallucinations are not as uncommon as most people would think. Researchers have found that 12 to 17 percent of children ages nine to twelve and 5.8 percent of adults hallucinate during the day.[11] Additionally, 37 percent of adults experience hallucinations when they are falling asleep, also known as "hypnagogic hallucinations."[12] Most of these people are not diagnosed with mental disorders.

Underactive Brain Functions

Underactive brain cells and brain networks have been documented in many mental disorders. This concept easily explains at least some of the symptoms that we see. I distinguish underactive function from the absence of function because underactive function implies the cells are still alive and able to work at least some of the time. This is important, as it means that symptoms will wax and wane. Sometimes things can seem normal, and other times the person can have symptoms. Here are some examples:

- People with ADHD can have a reduction in the activity of norepinephrine neurons in the locus coeruleus. These neurons help people focus, plan, and stay on task, so a reduction in their activity results in symptoms of ADHD.
- Cognitive problems, such as memory impairment, can be due to reduced function of the neurons that are involved in the storage and retrieval of memories. These are clearly affected in Alzheimer's disease but also in most of the chronic psychiatric disorders. People with chronic mental disorders often have cognitive impairment, even if this isn't part of their diagnostic criteria.
- At least one aspect of depression can involve a reduction in the activity of a brain system called the default mode network.[13] This results in a slowing or disorganization of normal brain function.
- "Emotional regulation" is a term used to describe symptoms in many different disorders, including mood, personality, and anxiety disorders. There are brain systems that are designed to help us control our emotional responses and regulate our moods. In some people, these areas of the brain appear to be underactive, resulting in symptoms such as unstable moods and anger outbursts.

Absence of Specific Brain Functions

Some mental disorders involve permanent changes in brain cells and connections. There are two primary causes of this—*developmental problems* and *cell death*. These problems are often associated with *neurodevelopmental* and *neurodegenerative* disorders, respectively. Cell death can also occur from things like a stroke or brain injury, which are different from neurodegenerative disorders, but these, too, can also result in mental symptoms.

There are many neurodevelopmental disorders. Autism is one example. Neurons and/or connections between neurons appear to be missing or at least different.

Neurodegenerative disorders, such as Alzheimer's disease, are associated with brain shrinkage and death of neurons. Once neurons die, there is usually no way to bring them back.

In both cases, cells or connections that are supposed to be present are not, so the brain is unable to perform these functions. Symptoms due to these permanent changes are always present. They don't wax and wane. The social deficits seen in autism are fixed. At least some of the cognitive deficits seen in Alzheimer's disease are also fixed. These don't change from day to day. However, both autism and Alzheimer's disease are also associated with ongoing mental symptoms that do wax and wane—anxiety, psychosis, and mood changes, to name a few.

These three scenarios—overactive, underactive, and absent brain functions—can account for all symptoms of mental disorders. However, there are two additional situations that are worth mentioning, because at first glance, they may not appear to fit neatly into these categories: multifaceted brain adaptations and behavioral disorders.

Multifaceted Brain Adaptations

The brain sometimes has complicated responses to situations that involve multiple symptoms, with some of them representing activation of some brain functions and inactivation of others. I will discuss depression, hypomania, and the trauma response. All of these can be normal and adaptive when they occur at the right times under the right circumstances. They are similar to the activation of the sympathetic and parasympathetic nervous systems, which involve a complicated array of brain and body functions, some that get turned on and others that get turned off.

Depression is a normal reaction to many stressors, adversities, and losses. Almost everyone has been depressed at least once. It usually doesn't last unrelentingly for two or more weeks, but it's a normal brain response. Although it commonly includes changes in mood, energy, appetite, and sleep, these changes can be very different in different people. Some people appear to have overactivity of the appetite system and others appear to have underactivity, resulting in eating too much or too little, respectively. Likewise, some people can sleep too much and others can't sleep enough. Distilling depression into distinct symptoms, some that represent overactive or underactive brain regions, is likely to be the most effective and accurate way to understand depression, even though depression often involves many symptoms.

Hypomania in many ways is the opposite of depression—people can feel great or euphoric, have increased energy, be more productive, and even get by on less sleep. This can also be normal. In fact, if it occurs in isolation, it's not a diagnosable disorder according to DSM-5. Most people have experienced symptoms of hypomania at some point or another in life. This commonly occurs when people fall in love, but it can also occur when people are excited about a project or an accomplishment, or when they have a spiritual awakening. Again, it may not last five or more days, but it can occur, which suggests that these brain functions are hardwired into all of our brains.

The trauma response is also normal. These symptoms include flashbacks and nightmares, avoidance of situations that remind the person of the event, negative effects on mood and thinking (similar to depression), trouble sleeping, being on edge and hyperalert, and other symptoms. One research group studied women shortly after being raped and found that 94 percent of them had these types of symptoms in the first weeks.[14] So, all of these responses can be "normal."

These multifaceted brain adaptations become *disorders* when they are *overactive*. They can be activated at the wrong time, last too long, or result in excessive or exaggerated symptoms. In some cases, they can be activated out of the blue for no clear reason—a hyperexcitable activation of the system. In other cases, they can be activated for a clear reason, such as a major stressor in life, but then they fail to inactivate after an appropriate amount of time. They become "stuck" in an "on" position when they should be turning off. This is similar to hyperexcitable pain cells in many pain disorders. Sometimes they can fire for no clear reason, but other times, the slightest injury or just moving the wrong way can trigger pain.

Behavioral Disorders

Some disorders are seen primarily as behavioral—in particular, substance use and eating disorders. These deserve special attention as well. Recall that they show strong bidirectional relationships with all mental disorders. I have said that mental disorders can be understood broadly as overactive, underactive, or absent brain functions. But these are behaviors . . . ones that people "choose" to engage in. What do these behaviors have to do with the brain malfunctioning?

There are three ways to think about it. The first is that eating and using addictive substances are behaviors that are controlled by our brains. There are clear pathways that control cravings, appetite, motivation, self-control, impulsivity, and novelty seeking. So, in some cases, if these parts of the brain are overactive or underactive, they might drive people to engage in these behaviors, which can lead to problems. The second possibility is that people may have symptoms of other mental disorders (due to overactive or underactive brain regions) and use alcohol, drugs, or change eating behaviors to cope with those symptoms. This is commonly referred to as the *self-medication hypothesis*. The third possibility is that some people can be perfectly fine and "normal" and begin to engage in these behaviors. Some can start using drugs or alcohol due to peer pressure alone. A person can start dieting due to peer pressure alone. As I will discuss later, all of these behaviors can have powerful effects on metabolism and the brain. They can lead to metabolic abnormalities, which can lead to overactivity and underactivity of specific brain functions, that can then trap people in vicious cycles—ones that we call eating disorders and substance use disorders.

A Complex Puzzle

One of the challenges in identifying what causes mental illness is that the findings I've mentioned here, such as reduced default-mode network activity leading to depression, are not consistent across people with the same disorder, or even in the same person at different times. Other than in cases of developmental abnormalities or cell death, the symptoms wax and wane, and so can the neuroscience findings. That's why we don't yet have diagnostic tests. The developmental abnormalities are not consistent for specific disorders, either. They can affect a wide range of cell types and brain regions in different people, even people diagnosed with the same disorder. When looking at brain changes and the way the brain functions in mental illness, heterogeneity and inconsistent findings are the name of the game.

That's a lot to account for. It's no wonder that the puzzle of mental illness has been so difficult to solve. What makes different parts of the brain overactive or underactive, leading to symptoms of mental illness? What makes the symptoms wax and wane? What exactly is causing these developmental abnormalities or areas of cell shrinkage and death? Why are they different in different people? All of these questions need to be addressed by any theory that tries to account for all mental disorders. I'm excited to share with you that the brain energy theory can do it. And it's all through *one common pathway*.

Chapter 7

Magnificent Mitochondria

I now return to the discussion of metabolism in order to continue to put the pieces of the puzzle together. Recall the traffic analogy from Chapter Five where each car was like a human cell. I described metabolism as very complicated. It's constantly changing, and may be different in different cells at different times. To be fair, that isn't really *one common pathway.* It's more like hundreds of different metabolic pathways.

But what controls metabolism? How do food and oxygen know where to go? What changes the metabolic rate in various cells? What makes some cells slow down, while others go faster? What is driving this intricate network of the human body?

Some would say it's the brain. Although the brain plays a critical role in metabolism, it can't control metabolism in all the different cells of the body at the correct times. Like city traffic, there must be some degree of control at the level of each car, or in the case of the human body, at the level of each cell. Cells receive input from other cells telling them to stop or go. Cells in close proximity also have signals that make neighboring cells stop or go (think brake lights on a car). But some of the signals are sent all over the body. They might originate in a brain cell or a liver cell, but they then travel long distances to affect cells throughout the body. These processes all lead to a coordination of metabolism, just like city traffic is coordinated on many levels, too.

There are many things that make city traffic flow. The different types of roads and highways. The different speed limits on different roads. The stop signs and stoplights. These are all important to the organization and flow of city traffic. But in the end, the true and primary force controlling the flow of traffic boils down to the drivers in the cars. They know the rules and they follow them. They make the cars go. They make the cars stop. They use turn signals. They look out for trouble. They steer the cars around problems. They drive the cars to their destinations. And even though the drivers don't know what's happening with all the other cars on all the other roads or highways, everything works.

Do human cells have "drivers" making the cells stop and go? It turns out that they do. The drivers of human cells, and human metabolism, are called *mitochondria*. And *they* are the *common pathway* to mental and metabolic disorders.

If you've ever taken a biology course, you likely remember that mitochondria are the "powerhouses of the cell." Mitochondria make energy for cells by turning food and oxygen into ATP. While there's no question that their role in energy production is critical, mitochondria are so much more than powerhouses. Without them, life as we know it wouldn't exist.

In the 2005 book *Power, Sex, Suicide: Mitochondria and the Meaning of Life*, Dr. Nick Lane provides a thorough and compelling story of mitochondria and their role in human evolution.[1] Although the title might suggest a pop culture quick read, Lane provides a rigorous scientific history of mitochondria and their role in human health and life itself.

Mitochondrial Origins

Once upon a time, the first mitochondrion (mitochondria is the plural of a single mitochondrion) was a bacterium. Researchers estimate mitochondria evolved from an independent living organism sometime between one and four billion years ago. A 1998 paper published in *Nature* suggests they share many genes with the modern-day *Rickettsia prowazekii*, a bacterium that causes typhus.[2] Billions of years ago, another single-cell organism, an *archaea*, engulfed this ancestral mitochondrion. Instead of the mitochondrion dying after being engulfed, as usually happens, they both lived. This new organism is thought to have evolved into the first *eukaryotic* cell (a cell with a nucleus). The inside bacterium began to focus on making energy, and the outside organism could focus on getting food. Make no mistake—this is important. It is not a trivial fact.

Thus, before there was a cell nucleus housing human DNA, and before there were other organelles, there was a mitochondrion—a single mitochondrion and a single host cell. Together, they were determined to survive. Actually, not just survive, but thrive. Like all forms of life—they were in it to win it. And win it they did!

Over time, it was this symbiotic arrangement that allowed for multicellular life—essentially all life that we can see with our eyes today. In all eukaryotes, these internal bacteria evolved into mitochondria. In plants and algae (also eukaryotes), some of them also evolved into what we now call *chloroplasts*. Although mitochondria and chloroplasts have different names, they look and function similarly, and they are thought to be descended from the same bacterium from billions of years ago. Furthermore, it's believed that this merger happened only once, and that all plants, animals, algae, and fungi that exist today descended from this same organism. For those who believe in God, this concept of a single event starting life as we know it might be reassuring. For those who don't believe in God, it was just one of those unusual and unlikely events that shaped evolution for billions of years to come. Regardless of what you believe, it was an important event in the story of life.

In evolution, being first matters. For example, when genes overlap among different organisms, they are usually believed to be more important than genes that are unique to specific species. The unique genes are thought to have occurred more recently in the evolutionary timeline, while the common genes developed much earlier. Things that have persisted for a long time are thought to be more essential to life. There are at least two reasons for this. The first is that evolution tends to get rid of things that are not essential or don't confer some advantage in terms of survival or reproduction. If organisms evolve to no longer need a trait, it will no longer be selected for and will often eventually disappear. The second is that new genes and traits must develop with and adapt to the genes and traits that are already there. Mitochondria were in eukaryotic cells first. Initially, it was just a single bacterium and a single outside cell. Over time, the nucleus and other organelles developed. As important as these other organelles are, mitochondria were there first. They likely influenced the development of these other cell parts and became indispensable. In fact, these other cell parts don't work correctly without mitochondria.

Modern Mitochondria

Mitochondria are no longer able to replicate themselves outside of a eukaryotic cell. In humans, mitochondria transferred most of their DNA to the cell's nucleus, where human DNA resides. There are about 1,500 mitochondrial genes that are now embedded within human DNA. These 1,500 genes make proteins that are required to either create or maintain mitochondria, and these proteins are shared with all the mitochondria in the cell. However, mitochondria didn't give up all their DNA. Each one of them still has thirty-seven genes. Individual mitochondria can use that DNA on their own—and thus mitochondria maintain some degree of independence, both from each other and from the cell in which they reside. This is highly unusual in biology, and its purpose is the subject of debate. The point, however, is this: Mitochondria and human cells are now 100 percent committed to each other. Neither can survive without the other.

Mitochondria are tiny. On average, each human cell has about three to four hundred mitochondria.[3] This means that there are about ten million billion mitochondria in the human body. They make up about 10 percent of our body weight despite their tiny size. In metabolically demanding cells—such as brain cells—a single cell can contain thousands of mitochondria, with mitochondria making up 40-plus percent of the cell volume.

Mitochondria are busy. Although small amounts of ATP can be produced without mitochondria through a process called *glycolysis,* mitochondria produce the lion's share of ATP, especially for brain cells. In the average human adult, they make about 9×10^{20} ATP molecules every second.[4] One group of researchers looked at brain cells using specialized imaging techniques and found that a single neuron in the human brain uses about 4.7 billion ATP molecules every second.[5] That's a lot of ATP!

Mitochondria move. This is a fairly recent discovery based on new techniques for studying living cells.[6] When a cell is dead under the microscope, nothing moves, so it's easy to see why researchers didn't think that mitochondria would be mobile. Other organelles typically are not. The finding that mitochondria actually move around living cells was highly unexpected. If you'd like to see a video of mitochondria moving, you can see them in the *PLOS Biology* article in the endnotes.[7] There are many other videos available online. There is a network of microtubules and filaments throughout the cell, often referred to as the *cytoskeleton,* that mitochondria use for their movement. There are many mechanisms involved, which are beyond our scope, but the point is simple—some mitochondria move around.[8] However, it appears that not all mitochondria are moving. Some stay in one place, while others move.

Why are they moving? Well, one reason is that they appear to go to the places in the cell where things are happening and where energy is needed. Energy needs to be produced in the right amount, in the right place, at the right time, and it goes through an unimaginably fast recycling process that involves mitochondria. The mitochondria that aren't moving appear to stay in places where things are always happening—either near factories where proteins are made (*ribosomes*) or synapses where there is a lot of activity, which is a very important fact relevant to how the brain functions. Researchers looking at brain cells under microscopes have known for decades how to identify where the synapses are—they look for the mitochondria.

Mitochondria are rapid recyclers. ATP is the energy currency of human cells. When it is used as energy, a phosphate group is removed, which turns it into adenosine diphosphate, or ADP. This ADP can't supply much energy anymore, but if a phosphate group is added back to it, it's as good as new. That's what mitochondria do. They take ADP and turn it back into ATP by attaching a phosphate group, then transfer it out to the cell cytoplasm where it is needed. They give one ATP and recycle one ADP simultaneously. If there is a lot of activity in a particular part of the cell, you will find mitochondria there. They have to provide the ATP, but they also suck up all of the ADP and recycle it. You can think about mitochondria as little vacuum cleaners, going around the cell and sucking up ADP and churning out ATP.

Remember that I said there were billions of ATP molecules being used every second in just one brain cell? Well, if there isn't a mitochondrion or two (well . . . maybe more than that) in the right place at the right time to both deliver all that ATP and recycle all the ADP, things will back up quickly and either slow down or stop working.

However, mitochondrial movement is more important than just making sure enough energy is supplied in the right place at the right time. It's also related to mitochondrial interactions with other organelles and with each other. These interactions are critically important to almost all cell functions and even gene expression.

To demonstrate the role of mitochondria, I'll first need to review some basic information on how neurons work. Although the function of any cell is complicated, and brain cells even more so, there are some basics that are directly regulated by mitochondria. Better understanding them will allow me to tie metabolism and mitochondria to distinct functions of brain cells. I'll use the next chapter to explain how all the symptoms of mental illness are directly related to mitochondria and metabolism.

Neurons have a resting membrane potential. Basically, this means that the inside of the cell has a negative charge compared to the outside of the cell. This charge is critically important to the function of the cell. It is created by ion pumps, which pump sodium, potassium, calcium, and other ions either inside or outside of the cell, or between compartments within the cell. These pumps all require energy.

Cells do a lot of ion pumping in order to set themselves up to be ready to fire. When the cell is triggered, it sets off a cascade of events that results in the cell doing its thing, whether that entails releasing a neurotransmitter or a hormone, or doing something else. It's like setting up a row of dominoes. It takes time and work to set them up, but it's easy to push them all over by simply nudging one of them. Once they all fall down, they need to be set up again. That requires more work. Mitochondria provide almost all the energy needed to do all of this.

What Else Do Mitochondria Do?

Calcium levels play an important role in the function of cells. High levels of calcium in the cytoplasm can trigger all sorts of things to happen. In many ways, calcium is an "on/off" switch. When levels are high, the cell is "on." When levels are low, the cell gets turned "off." Mitochondria are directly involved in calcium regulation. When mitochondria are prevented from functioning properly, calcium regulation is disrupted—and this important "off" switch can be as well.[9] Therefore, mitochondria are essential in turning cells both on and off. They provide the energy needed for ion pumping, and they also regulate the calcium levels that function as essential on/off signals.

Energy and mitochondria are required to turn cells both on *and* off. This may seem paradoxical, but it will make more sense if you think of the "off" switch as electronic brakes on a car that require energy to work. Without enough energy to apply the brakes fully and quickly at appropriate times, the car can become impossible to control and cause major disruptions in the flow of traffic. These dichotomous consequences of metabolic and mitochondrial dysfunction are important to understand. Some cells will stay on too long when they are energy deprived, while other cells will fail to work. I will come back to this soon enough.

Turning cells on and off is critically important. Understanding this function will help us explain most of the symptoms of mental illness. However, mitochondria actually do much more than that. Their role in human health is a cutting-edge, vigorous area of research that spans almost every field in medicine.

Let's outline some of the other roles that mitochondria play that are important to their relationship with mental health.

Mitochondria Help Regulate Metabolism Broadly

In 2001, a peptide called *humanin* was first reported to have broad effects on metabolism and health.[10] The gene for this peptide appears to reside on both mitochondrial DNA and nuclear DNA. It was first discovered in research on Alzheimer's disease. Since its discovery, two other peptides, MOTS-c and SHLP1–6, have been discovered and added to a new class of molecules called *mitochondrially derived peptides*. The genes for these peptides are on mitochondrial DNA, and these peptides are produced by mitochondria. They are now of great interest to researchers. They have been shown to have beneficial effects on illnesses such as Alzheimer's disease, strokes, diabetes, heart attacks, and certain types of cancer. They also have broad effects on metabolism, cell survival, and inflammation.[11] The existence of these peptides suggests that mitochondria are able to communicate with each other through these peptide signals in order to regulate metabolism throughout the body.

Mitochondria Help Produce and Regulate Neurotransmitters

Neurotransmitters have been a primary focus in the mental health field. It turns out that mitochondria play critical roles in their production, secretion, and overall regulation.

Neurons often have one specific neurotransmitter that they specialize in making. Some make serotonin. Others make dopamine. The process of making a neurotransmitter takes energy and building blocks.

Mitochondria provide *both*. They play a direct role in the production of acetylcholine, glutamate, norepinephrine, dopamine, GABA, and serotonin.[12] Once made, neurotransmitters are stored in vesicles, or little bubbles, until they are ready to use. Vesicles filled with neurotransmitters travel down the axon to get to their ultimate release site. That takes energy. The signal to release neurotransmitters depends upon the resting membrane potential and calcium levels that I discussed. Once that signal comes, the actual release of neurotransmitters also takes energy. Fascinatingly, once neurotransmitters are released at one location, the mitochondria move to another location of the cell membrane to release a new batch of neurotransmitters.[13] Once released, neurotransmitters have their effect on the target tissue, whether it's another nerve, muscle, or gland cell. After they are released from the receptors on the target cell, they are sucked back into the axon terminals (a process called *reuptake*), and you guessed it, that takes energy. They are then repackaged back into vesicles for the next round—yet more energy.

Mitochondria are normally found in large supply at synapses. When they are prevented from getting to the synapses, neurotransmitters don't get released, even if there is ATP present.[14] When mitochondria aren't functioning properly, neurotransmitters can become imbalanced. Given that neurotransmitters are an important way for nerve cells to communicate with each other, imbalances can disrupt normal brain functions.

The role of mitochondria in regulating neurotransmitters goes much further than just their involvement in synthesis, release, and reuptake. Mitochondria actually have receptors for some neurotransmitters, indicating a feedback cycle between neurotransmitters and mitochondria. They also have some of the enzymes involved in the breakdown of neurotransmitters, such as monoamine oxidase. They are involved in regulating the release of GABA, and they actually store GABA within themselves.[15] Finally, several neurotransmitters are known to regulate mitochondrial function, production, and growth. Clearly, neurotransmitters are much more than just messengers between cells impacting mood. They are essential regulators of metabolism and mitochondria themselves. I'll come back to this later.

Mitochondria Help Regulate Immune System Function

Mitochondria also play an essential role in immune system function.[16] This includes fighting off viruses and bacteria, but it also includes low-grade inflammation, something that has been found in most metabolic and mental disorders to some degree. Mitochondria help regulate how immune cells engage with immune receptors. When cells are highly stressed, they often release components of mitochondria, which serve as a danger signal to the rest of the body, one that activates chronic, low-grade inflammation.[17]

One study looked at specific types of immune cells called macrophages to see how these cells coordinate the complicated repair processes in wound healing. The cells do different things during different phases of healing. Up until this study, it wasn't known how the cells know when and how to change between phases. The researchers found that mitochondria specifically controlled these processes.[18]

Mitochondria Help Regulate Stress Responses

We now know that mitochondria help control and coordinate the stress response in the human body. This includes both physical and mental stressors. Physical stressors include things like starvation, infection, or a lack of oxygen. Mental stressors are anything that threatens or challenges us (as talked about in the previous chapter).

When cells are physically stressed, they initiate a process called the *integrated stress response.* This is a coordinated effort by the cell to adapt to and survive adverse circumstances through changes in metabolism, gene expression, and other adaptations. Many lines of research show that mitochondrial stress itself leads to the integrated stress response.[19] If the cell isn't able to manage the stress, one of two things happens—it either triggers its own death, a process called *apoptosis,* or it enters into a "zombielike" state called *senescence,* which has been associated with aging and many health problems, such as cancer.

Up until recently, it wasn't known how the different aspects of the psychological stress response are all coordinated in the body and brain. It turns out that mitochondria play a critically important role! One brilliant study by Dr. Martin Picard and colleagues demonstrated this, and its title says it all: "Mitochondrial functions modulate neuroendocrine, metabolic, inflammatory, and transcriptional responses to acute psychological stress."[20] These researchers were studying mice and genetically manipulated their mitochondria to see what effects these manipulations had on the stress response. They manipulated only four different genes—two located in mitochondria themselves and two located in the cell nucleus that code for proteins used exclusively in mitochondria. Each genetic manipulation resulted in different problems with mitochondrial function. However, even with only four manipulations, they found that all the stress response factors were affected. This included changes in cortisol levels, the sympathetic nervous system, adrenaline levels, inflammation, markers of metabolism, and gene expression in the hippocampus. Their conclusion was that mitochondria are directly involved in controlling all these stress responses, and if mitochondria aren't functioning properly, these stress responses are altered.

Mitochondria Are Involved in Making, Releasing, and Responding to Hormones

Mitochondria are key regulators of hormones. Cells that make hormones require more energy than most. They synthesize the hormones, package them up, and release them, just as I described for neurotransmitters. It takes a lot of ATP to do this, and mitochondria are there to deliver it.

For some hormones, mitochondria are even more important—this includes well-known names like cortisol, estrogen, and testosterone. The enzymes required for initiating the production of these hormones are found only in mitochondria. Without mitochondria, these hormones aren't made. But there's more. Mitochondria in other cells sometimes have receptors for these hormones. So, in some cases, these hormones can begin in mitochondria in one type of cell and end with mitochondria in another type of cell.

Mitochondria Create Reactive Oxygen Species (ROS) and Help Clean It Up

Mitochondria burn fuel—either carbohydrates, fats, or protein. Burning fuel can sometimes create waste products. When mitochondria burn fuel, electrons flow along the electron transport chain. These electrons are a source of energy usually used to make either ATP or heat. However, sometimes these electrons leak outside of the usual system. When they do, they form what are called *reactive oxygen species* (ROS).[21] These include molecules such as superoxide anion (O_2^-), hydrogen peroxide (H_2O_2), hydroxyl radical (•OH), and organic peroxides. At one point, researchers believed ROS were simply toxic waste products. We now know small amounts of ROS actually serve a useful signaling process inside the cell. For example, a 2016 paper published in *Nature* found that ROS were the primary regulators of heat production and energy expenditure—a broad measure of metabolic rate.[22] However, large amounts of ROS *are* toxic and result in inflammation.[23] You may have heard the term *oxidative stress*—that's what this is! ROS are known to cause damage to mitochondria and cells. They are associated with aging and many diseases. Given that ROS are produced right in the mitochondria and are highly reactive, they often damage mitochondria first. The mitochondrial DNA is unprotected, so large amounts of ROS are known to result in mitochondrial DNA mutations. These ROS can also damage the mitochondrial machinery itself. If they leak outside of the mitochondria, they can damage many different parts of the cell.

Additionally, mitochondria serve as ROS janitors. As well as producing ROS, mitochondria also clean up some of it up through an elaborate system of enzymes and other factors that serve to detoxify ROS.[24] Cells have other antioxidant systems, too, but mitochondria play a role. When this detoxification system fails, these ROS waste products can pile up and cause damage. This can lead to cellular dysfunction, otherwise known as aging, cell death, and disease.

Mitochondria Are Shape-Shifters

Mitochondria change shape in response to different environmental factors. Sometimes they are long and thin. Other times they are short and fat. Sometimes they are round. In addition to changing shape, they interact with each other in profound ways. They can merge to make just one mitochondrion—a process called *fusion*. They can divide and form two mitochondria—a process called *fission*. These changes in shape are very important to cell function. In 2013, two articles published in the journal *Cell* showed that the process of mitochondria fusing with each other significantly impacts fat storage, eating behaviors, and obesity.[25] Mitochondrial changes in shape and their fusion with each other appear to create signals that can affect the entire human body. When mitochondria are prevented from doing these things, metabolic problems ensue, not just in the cells affected, but sometimes throughout the body.

Mitochondria Play a Primary Role in Gene Expression

Nuclear DNA is where the human genome resides. It's contained within the cell nucleus. Researchers once thought that genes controlled everything about the human body. They assumed that the nucleus was the control center of the cell. We now know that it's not always about the genes themselves, but more about what causes certain genes to turn on or off. This is the field of *epigenetics*.

Mitochondria are primary regulators of epigenetics. They send signals to the nuclear DNA in several different ways. This is sometimes referred to as the *retrograde response*.

It has long been known that the ratio of ATP to ADP, levels of ROS, and calcium levels can all affect gene expression. As you now know, these are all directly related to mitochondrial function. However, given that these are also markers of general cellular health and function, no one thought too much of it. They certainly didn't think of it as a way for mitochondria to directly control the expression of genes in the nucleus.

In 2002, it was discovered that mitochondria are required for the transport of an important epigenetic factor, nuclear protein histone H1.[26] This protein helps regulate gene expression and is transported from the cytoplasm to the nucleus, a process that requires ATP. Researchers discovered, however, that ATP alone isn't enough. Mitochondria must be present in order for this transfer to occur. Without mitochondria, this transfer doesn't happen.

In 2013, it was discovered that mitochondrial ROS directly inactivate an enzyme called histone demethylase Rph1p, which regulates epigenetic gene expression in the cell nucleus.[27] This process was found to play a role in extending lifespan in yeast and is thought to possibly play a role in humans as well.

In 2018, two additional studies demonstrated even more of a role for mitochondria in gene expression. The first was a report by molecular biologist Maria Dafne Cardamone and colleagues showing that a protein, GPS2, is released by mitochondria in response to metabolic stress.[28] Metabolic stress can be caused by a lot of different things, but starvation is a clear example. After GPS2 is released by mitochondria, it enters the cell nucleus and regulates a number of genes related to mitochondrial biogenesis and metabolic stress.

Another group of researchers, Dr. Kyung Hwa Kim and colleagues, found another mitochondrial protein, MOTS-c, that is coded for by mitochondrial DNA and plays a role in gene expression.[29] This was very unexpected. Up until about twenty years ago, everyone assumed that mitochondrial DNA was just about machinery needed for ATP production. MOTS-c gets produced in response to metabolic stress as well. After MOTS-c is produced in the mitochondria, it makes its way into the nucleus and binds to the nuclear DNA. This results in the regulation of a broad range of genes—ones related to stress responses, metabolism, and antioxidant effects.

Finally, and most spectacularly, Dr. Martin Picard and colleagues experimentally manipulated the number of mitochondria with mutations in cells and found that as they increased the number of dysfunctional mitochondria, more epigenetic problems and changes occurred.[30] The impact was on almost all of the genes expressed in the cells. Ultimately, in situations in which almost all the mitochondria were dysfunctional, the cells died. This study provided evidence that mitochondria are not just involved in the expression of genes related to energy metabolism, but possibly in the expression of *all* genes.

Mitochondria Can Multiply

Under the right circumstances, cells will make more mitochondria—a process called *mitochondrial biogenesis*. Some cells end up with a lot of mitochondria. These cells can produce more energy and function at a higher capacity. It is widely believed that the greater the number of healthy mitochondria in a cell, the healthier the cell. We know that the number of mitochondria decreases with age. We also know that the number of mitochondria decreases with many diseases. People who are considered the "fittest" among us—athletic champions—have more mitochondria than most, and their mitochondria appear to be healthier.

Mitochondria Are Involved in Cell Growth and Differentiation

Cell growth and differentiation is a complicated process during which a generic stem cell becomes a specialized cell. *Differentiation* means that the cells become different from each other and take on specialized roles. Some become heart cells. Others become brain cells. Within the brain, different cells take on varying roles. Brain cells change throughout life. Some form new synapses. Some prune unnecessary parts. Some grow and expand when needed. This is *neuroplasticity*.

This process of growth and differentiation involves activation of specific genes in the right cells at the right times. It also involves many signaling pathways. Lastly, it involves the production of building blocks for new cells and new cell parts, balanced with energy needs.

It has long been known that mitochondria are essential to cell growth and differentiation. Most researchers assumed it was simply a matter of their powerhouse function since cell growth and differentiation require energy. Recent research, however, strongly suggests a much more active role. Their regulation of calcium levels and other signaling pathways are essential to this process.[31] Their fusion with each other appears to send signals that activate genes in the nucleus. When mitochondria are prevented from fusing with each other, the cells don't develop correctly.[32] Other research has shown that mitochondrial growth and maturation is essential to proper cell differentiation.[33] Still other research has shown a direct and essential role of mitochondria in the development of brain cells.[34] The bottom line is that cells don't develop normally when mitochondria aren't functioning properly.

Mitochondria Help Maintain Existing Cells

In the previous chapter, I discussed *autophagy* and cell maintenance. It turns out that mitochondria are directly involved in this process, too. They generate many of the signals, such as ROS and other metabolic factors, that play a key role in autophagy. They also interact with other parts of the cell, such as lysosomes, that are involved in the process. Maintenance work takes energy and building blocks as well, and mitochondria are there to provide both.

Mitochondria appear to be in a complicated feedback cycle with autophagy, as dysfunctional mitochondria can be removed and replaced with healthy mitochondria in a process known as *mitophagy*. Mitochondria can be beneficiaries of autophagy, but they also play a role in stimulating autophagy more broadly for the entire cell.[35]

Mitochondria Eliminate Old and Damaged Cells

Cells die every day. There are two well-known types of cell death—*necrosis* and *apoptosis*. Necrosis occurs when a cell is abruptly killed, such as a heart cell dying during a heart attack. Necrosis is a bad thing. Apoptosis occurs when cells get old or damaged. Apoptosis is a planned process often referred to as *programmed cell death*—the signal to die actually comes from the cell itself. Overall, apoptosis is seen as an extraordinarily good thing for human health and survival. It allows old cells to be replaced by new ones. It eliminates damaged cells that might turn into cancer. Every day, about ten billion cells in the human body die and are replaced by new ones.[36]

It was once thought that genes in the nucleus controlled apoptosis. We now know that's not true. It's mitochondria. When mitochondria experience high levels of stress and accumulate large amounts of ROS, they begin to degrade. When this happens, they release a protein called *cytochrome c*, which then activates what are called "killing enzymes"—the *caspases*. These enzymes degrade everything in the cell until it dies. Many of the cell parts get recycled.

Autophagy and apoptosis are somewhat related, but they are different processes. Autophagy is usually about repairing and replacing parts within a cell, but the cell usually remains alive. Apoptosis is the death of an entire cell. Nonetheless, they are both required for health and longevity, and mitochondria play a role in both.

There are even more types of cell death, beyond the scope of this book. Nonetheless, one review was able to link all of them to the functions of mitochondria.[37]

Putting It All Together

Change is hard. Models, practices, and conceptual frameworks are difficult to shift. But what if our ideas about the control of cells have been all wrong?

If we go back to our automotive analogy, I suggested that each cell was like a car in the congested traffic of a large city. If we look inside that car, there are many drivers—all the mitochondria. It might be easier to change the metaphor at this level and think of the inside of each cell as a factory. The factory receives supplies, such as glucose, amino acids, and oxygen, and performs a function. Some make neurotransmitters. Others make hormones. Some are muscle cells and cause the body to move. Mitochondria are the workers inside those factories (mitochondria as workers is an analogy Nick Lane used in his book as well).[38] There are many different roles and tasks for them. Some mitochondria help with the production and release of hormones or neurotransmitters. Others serve as janitors—helping to clean up ROS and other debris. Some help communicate with the nucleus—sending signals to turn genes on or off. They are the regulators of calcium, ROS, and other important signals in cells. They work together and communicate with each other—they fuse with each other, move around cells, and communicate with mitochondria in other cells through hormones, such as cortisol, and through other mechanisms, such as mitochondrially derived peptides. And of course, they provide most of the power—or ATP—to make the factory work. When workers in one cell aren't doing well, they not only affect the rest of the workforce in that cell, but they can also affect the workers in other cells.

Over the past twenty years, a lot of the new evidence on the role of mitochondria in the cell has been shocking and unexpected. Almost no one thought that mitochondria could control the regulation of genes in the nucleus—both on a daily basis and during cell growth and differentiation. Their interaction with and regulation of other organelles, such as the endoplasmic reticulum and lysosomes, were also surprising. They were usually thought of as relatively insignificant, and very small, ATP factories. They were sometimes described as "little batteries." Many researchers still see them this way.

For centuries, researchers have been trying to figure out how cells work. Up until recently, they have focused primarily on all the big parts of cells and largely ignored the tiny little mitochondria. Many still think the nucleus, with its coveted human genome, is the control center. Others think that it's all about the outside cell membrane and the different receptors that are embedded into it. Different neurotransmitters or hormones make cells do things. What if both takes have some truth in them, but the real story is about the mitochondria—the workers? Given all the roles that mitochondria play in so many different aspects of cell function, is it possible that they are the real answer to understanding how cells work? What if all the different organelles in a cell are just big machines or storage sites to be used by mitochondria to do different tasks in a cell? Could the nucleus simply be a large storage center for the DNA, the blueprints of the cell, to be used when called for by the mitochondria? Could the other organelles be large machines, ones that make proteins (ribosomes) or waste disposal machines (lysosomes), to be used by the mitochondria for these different purposes? After all, mitochondria are the only organelles that move around the cell, interact with each other, and interact with all the other organelles. Mitochondria were in the cell first. They were the first organelle. They were also an independent living organism at one point. In many ways, the evidence can't rule this out.

To be clear, I'm not suggesting that mitochondria have brains and make independent decisions on all of these functions. Instead, I am suggesting that they are like little robotic workers, doing what they are programmed to do. They are longtime, loyal servants to human cells. But like so many unappreciated servants and workers, maybe they deserve a little more respect and recognition for all that they do.

Whether you like this analogy or not, even if you want to continue to think of mitochondria as nothing more than little batteries, one thing is abundantly clear and uncontroversial—when mitochondria aren't working, neither is the human body or brain.

Chapter 8

A Brain Energy Imbalance

In the previous chapter, I reviewed mitochondria in all their glory. Their function affects every cell in the human body. Their involvement in all aspects of cell function, neurotransmitters, hormones, inflammation, immune system function, regulation of gene expression, development, and the maintenance and health of cells results in widespread effects throughout the body and the brain. They are the drivers of cells and metabolism. They are the workforce of the human body.

But the question remains: Do we have evidence that metabolic problems are related to mental disorders? And how?

Yes! There is an abundance of evidence that links problems with metabolism to mental disorders.

As I discussed in Chapter Five, physicians and researchers have known for well over a century that mental disorders appear to be linked to metabolic disorders, such as diabetes. Direct evidence of metabolic abnormalities in people with mental disorders, even those who don't yet have obesity, diabetes, or cardiovascular disease, dates back to at least the 1950s. Among the abnormalities found in markers of metabolism: differences in levels of

ATP, redox markers (the balance between oxidants, such as ROS, and antioxidants), hormones, neurotransmitters, and lactate (a marker of metabolic stress). In the 1980s, it was discovered that infusing lactate into the vein of a person with panic disorder would often precipitate an immediate panic attack.[1] As I already discussed, cortisol dysregulation also seems to play a role, at least in some people, and this is a metabolic hormone.

Neuroimaging studies have provided an overwhelming amount of evidence for metabolic differences in the brains of people with mental disorders. Functional magnetic resonance imaging (fMRI) and near-infrared spectroscopic imaging (NIRSI) can measure localized changes in cerebral blood flow related to neural activity, which is an indirect marker of metabolism and brain activity. Positron emission tomography (PET), blood-oxygen-level-dependent imaging (BOLD), and single-photon emission computed tomography (SPECT) all measure some metric of metabolism—levels of glucose, oxygen, or a radioactive molecule that researchers inject into a person's vein. All these imaging studies are measuring metabolism in the brain because metabolism is a marker of brain activity. When neurons are active, they use more energy. When they are resting, they use less.

These studies have provided us with a plethora of data demonstrating differences in the brains of people with mental disorders compared to the brains of healthy controls. Some brain regions are overactive, while other brain regions are underactive. More recently, researchers have turned to *functional brain connectivity* studies, which look at the interactions of two or more brain regions in attempts to determine which of those regions communicate with each other to perform specific tasks. However, even with all this research, heterogeneity and inconsistent findings have been the name of the game. If you don't believe me, the American Psychiatric Association published a "Resource Document on Neuroimaging" in 2018 in which they concluded: "There are currently no brain imaging biomarkers that are clinically useful for any diagnostic category in psychiatry."[2]

The researchers doing this neuroimaging work, however, have known for decades that there are differences in metabolism in the brains of people with mental disorders. At first glance, they may think that the theory of brain energy offers nothing new. "Obviously mental disorders are related to metabolism! We've known that all along! Metabolism is everything in biology. What's new here?"

As I hope you'll come to understand, there is something new here. And not just new, but revolutionary. While these researchers have been lost in the overwhelming complexity of metabolism and how the brain works, trying to figure out what is making some brain regions overactive and other brain regions underactive, they have failed to see the big picture of metabolism. Most importantly, they have failed to see the role of mitochondria in all of it. By stepping back and looking at the bigger picture (even if that bigger picture is playing out on a microscopic level), we can find new ways to understand what's happening with metabolism and mental health and see new ways to address these problems.

Mitochondrial Dysfunction and Mental Health

But do we have evidence that mitochondria aren't functioning properly in people with mental disorders?

Yes! We now have an abundance of evidence.

Over the past several decades, it has become clear that mitochondria play a much larger role in human health than ever imagined. When mitochondria don't function properly, the human body doesn't function properly. *Mitochondrial dysfunction* is the term most frequently used to describe impairment in mitochondrial function. The diseases and illnesses that have been associated with mitochondrial dysfunction are widespread, and the list includes *almost all* of the psychiatric disorders. It also includes the metabolic and neurological disorders that I have discussed—obesity, diabetes, cardiovascular disease, Alzheimer's disease, and epilepsy. In reality, it includes even more disorders—many cancers and Parkinson's disease are among them. I won't be able to go into detail on all these different disorders. However, the framework that I am creating will apply to them as well.

The specific psychiatric disorders in which mitochondrial dysfunction has been identified include the following: schizophrenia, schizoaffective disorder, bipolar disorder, major depression, autism, anxiety disorders, obsessive-compulsive disorder, posttraumatic stress disorder, attention deficit/hyperactivity disorder, anorexia nervosa, alcohol use disorder (aka alcoholism), marijuana use disorder, opioid use disorder, and borderline personality disorder. Dementia and delirium, often thought of as neurological illnesses, are also included.

This list does not include every psychiatric diagnosis in DSM-5. However, that's not necessarily because mitochondrial dysfunction does not exist in the other diagnoses; research just hasn't been done on the other diagnoses yet. Nonetheless, this list is certainly broad enough to state that mitochondrial dysfunction has been found in a wide range of diagnoses that include pretty much every symptom found in psychiatry.

If all this evidence has existed for a while, why hasn't anyone else suggested that mitochondrial dysfunction is the common pathway to metabolic or mental disorders?

Well . . . they have! To most people reading this book, this may seem like new information. However, this book is not the first to assert the importance of mitochondria in human health and disease.

Dr. Raymond Pearl published a book on the *rate of living theory* in 1928, in which he argued that longevity and diseases of aging, which include most of the metabolic diseases, are due to metabolic rate. In 1954, Dr. Denham Harman proposed the *free radical theory of aging*, which focused on ROS as the cause of age-related diseases. In 1972, he further developed this theory and proposed the *mitochondrial theory of aging*, which recognized the central role of mitochondria in the production of ROS. In recent years, there has been an explosion of research on mitochondria and their relationship to obesity, diabetes, cardiovascular disease, and aging itself, with tens of thousands of research articles published in the medical literature.

The psychiatric literature is filled with articles from respected scientists highlighting the role of mitochondria in mental disorders. A 2021 medical literature search came up with more than four hundred articles relating to schizophrenia and bipolar disorder, more than three thousand relating to depression, more than four thousand relating to Alzheimer's disease, and more than eleven thousand relating to alcohol use. Some of this pioneering research has been close to home for me, coming from respected and internationally renowned colleagues, such as Professors Bruce Cohen and Dost Öngür at McLean Hospital and Harvard Medical School, where I have worked for more than twenty-five years.

In 2017, Dr. Douglas Wallace, the founder of the field of mitochondrial genetics, published an article in *JAMA Psychiatry*, one of the leading psychiatric journals, audaciously claiming (as I do in this book) that all psychiatric disorders are the result of mitochondrial dysfunction.[3] Being a geneticist, Wallace focused on mitochondrial genes. They mutate often due to ROS and the lack of protection of mitochondrial DNA. Wallace argued that the brain is the organ that will be most affected by a problem with energy production by mitochondria. He argued that different parts of the brain might fail first—likely because they are more sensitive to energy deprivation than others. This makes sense, given that most machines do have "weakest links." The brain is likely the same. So, a small amount of energy deprivation might result in ADHD or depression and a larger amount of energy deprivation might result in other disorders, such as schizophrenia.

The rebuttal was swift. Dr. Tamas Kozicz and colleagues argued that although people like "simple" explanations, psychiatric disorders do not lend themselves to such a simple explanation.[4] They acknowledged that "suboptimal mitochondrial function" does appear to play a role in most psychiatric disorders. However, focusing solely on mitochondrial energy production cannot account for the diversity of symptoms that we see in the billions of people with mental disorders. Furthermore, it can't even account for the diversity of symptoms that we see in people with rare genetic mitochondrial diseases. Even people with the same mitochondrial genetic mutation can have different symptoms. They argued that mental disorders are far too complex and different from person to person to be explained by a single factor.

What these researchers failed to consider is how many other roles mitochondria play in cells apart from the production of energy. They also failed to recognize just how many different factors affect the function and health of mitochondria. When mitochondria don't function properly, neither does the brain. When brain metabolism is not properly controlled, the brain doesn't work properly. Symptoms can be highly variable, but mitochondrial dysfunction is both *necessary and sufficient* to explain all the symptoms of mental illness.

Defining This Root Cause

As I just discussed in the previous chapter, mitochondria do a lot of different things. Defining what dysfunction means is difficult and has been a challenge to scientists; it can mean very different things in different research studies.

The same can be said for cars. If a car is "dysfunctional," what does that mean? It could mean that the engine sputters when traveling down the highway. It could mean that a tire is flat, and the car can't move along the road as easily. It could mean that the lights and the turn signals aren't working. Those are all different problems with a car. They all result from different causes. But here's an important point: Regardless of what is wrong with the car, if it's on the highway with any of those problems, it will affect the other cars on the road. It will be more likely to slow down traffic or cause an accident. Traffic can slow or completely stop. The highway could "stop working" due to one car. In reality, the overwhelming majority of car accidents aren't about the cars themselves but about the drivers of the cars. Drivers can also be "dysfunctional." They can be on their cell phones. They can fall asleep at the wheel. They can be drunk. They can be high. They can have road rage. Regardless of the cause of the dysfunction, whether it's the car or the driver, these wayward cars and drivers affect traffic in similar ways.

Mitochondrial dysfunction is the same. It can be caused by many different things and can result in different problems for the mitochondria and for the cells in which they reside.[5] Measuring the function of mitochondria is difficult. Remember how tiny they are? There are usually hundreds and sometimes thousands of them in one cell. Cells are pretty tiny on their own.

Mitochondrial dysfunction can stem from problems with the mitochondria themselves. This includes genetic mutations or a shortage of mitochondria in the cell. As I mentioned, mitochondria have their own DNA. It is not protected in the same way that the human genome is protected; therefore, it's prone to mutations. Mitochondria are producing ROS, and if they produce too much of it, it can damage the mitochondrial DNA or other parts. This can lead to defective mitochondria. When mitochondria are defective, they are supposed to be disposed of and recycled, with new ones taking their place. If this doesn't happen, a cell can be left with a workforce shortage. It's well established that the number of mitochondria in our cells decreases as we get older, resulting in less metabolic capacity for cells.

When the workforce is decreased, whether through aging or malfunctioning mitochondria, productivity goes down. As the mitochondria continue to decline, the cell usually dies. This leads to organs and tissues shrinking. As the cells die off, the organs get weaker and more vulnerable to stress. Brains shrink. People lose muscle mass. The heart isn't as strong. This phenomenon is also seen in people with chronic mental disorders. As I already mentioned, accelerated aging has been found in people with *all* mental disorders.

The more important cause of mitochondrial impairment is one that I'll call *mitochondrial dysregulation.* Many of the factors that affect mitochondrial function come from outside the cell. They include neurotransmitters, hormones, peptides, inflammatory signals, and even something like alcohol. Yep! Alcohol affects the function of mitochondria. I refer to this as dysregulation, as opposed to dysfunction, because in some cases, the mitochondria were functioning just fine, but their environment became hostile quickly and caused impairment—similar to people doing the best they can when they are highly stressed.

Defining which mitochondrial functions to focus on is critically important and varies in different studies. Some of these studies are done while mitochondria are still in living cells (*in vivo*), and others are done with mitochondria laid bare in a laboratory dish (*in vitro*).[6]

Many researchers have focused on ATP production. They can measure the amount of ATP relative to ADP in the cell cytoplasm and make assumptions about how well the mitochondria are functioning. The implications of this research are straightforward; ATP provides energy for the cell to work. If the levels are reduced, the cell won't work as well. The levels of ATP to ADP are also important signals within cells. This ratio affects numerous aspects of cell function, including gene expression. Reduced levels of ATP have been found in a wide variety of disorders, including schizophrenia, bipolar disorder, major depression, alcoholism, PTSD, autism, OCD, Alzheimer's disease, epilepsy, cardiovascular disease, type 2 diabetes, and obesity. Even though most people think of obesity as a surplus of energy, many cells in the bodies and brains of people with obesity are actually deprived of ATP, due to mitochondrial dysfunction.[7]

Other researchers have focused on *oxidative stress*. Remember, this is a term used to describe the buildup of ROS. Recall that mitochondria create ROS, but they also help detoxify them through antioxidants. When mitochondria aren't working properly, ROS build up and can cause damage to the cell in general, but more often to the mitochondria themselves, leading to a vicious feedback cycle. Numerous studies have found higher levels of oxidative stress in essentially all of the metabolic, neurological, and mental disorders that I have been discussing. This has been linked to cell damage and accelerated aging.

There have been three major shortfalls in the research on the role of mitochondria in health and disease to date:

1. **Focus on just one function.** Most studies have focused on only one function or aspect of mitochondria. They often fail to consider all the diverse functions. Some mitochondrial functions can be normal, while others are abnormal. Additionally, some functions can affect other functions. For example, studies that have looked at mitochondrial ATP production often view this as the primary, and sometimes only, role of mitochondria. Any adverse outcomes that they see in the cells they are studying are attributed to this

impaired ATP production. In reality, mitochondria that aren't producing enough ATP may also have trouble fusing with each other, or might be leaking large amounts of ROS, or might have trouble managing calcium levels in the cell. These functions may be more important to the cell deficits observed by the researchers, even if those functions weren't measured. In some cases, ATP production might be normal, while these other functions are abnormal, and the researchers might conclude that the mitochondria were functioning just fine when, in fact, they weren't.

2. **Differences between cells.** Mitochondria are affected by numerous factors, both within the cell and from outside the cell. The number and health of mitochondria are not equally distributed among all cells in the body and brain. Some cells can have perfectly healthy mitochondria in abundance, while other cells can have defective or insufficient mitochondria. Researchers must study specific cells to determine whether mitochondria in those cells are playing a role in an illness. Studying a healthy immune cell may not offer any information about what's happening in a malfunctioning brain cell.

3. **The role of feedback loops.** The question of which comes first, the chicken or the egg, has led many researchers astray. Does mitochondrial dysfunction cause illness? Or does illness cause mitochondrial dysfunction? Are mitochondria simply innocent bystanders and the victims of some other destructive process?

   When thinking about causes and consequences, things can get confusing. There are many causes of mitochondrial dysfunction. There are also many consequences. I'll get to this soon enough. What's confusing, though, is that *causes can lead to consequences, but consequences can lead to causes.* When we see this type of a pattern, we need to think about feedback loops. When it comes to metabolism and mitochondria, almost all things are regulated in a feedback loop.

   One example is in Alzheimer's disease research. We know that the abnormal protein *beta-amyloid* accumulates in the brains of

> people with Alzheimer's disease. This protein has been the primary target of research. We know the more beta-amyloid that is present, the more likely it is that someone will develop Alzheimer's disease. We also know it is toxic to mitochondria and causes mitochondrial dysfunction.[8] Many researchers have stopped there. They feel they have enough evidence that mitochondria are innocent bystanders of this destructive protein. What causes beta-amyloid to accumulate? They don't know. They are still looking for that cause. What they have missed, however, is that mitochondrial dysfunction might very well be the cause of the accumulation of beta-amyloid itself. We have evidence that mitochondrial dysfunction begins even before beta-amyloid starts accumulating.[9] It's possible that it's a positive feedback loop. Mitochondrial dysfunction causes a maintenance problem in cells. This results in the accumulation of beta-amyloid (a protein that should be getting disposed of and recycled). This beta-amyloid accumulation then makes mitochondrial dysfunction even worse. This feedback loop results in the downward spiral that we call Alzheimer's disease.

Fortunately, research in the last twenty years has broadened substantially regarding the scope of mitochondrial function—all the different roles mentioned in the prior chapter are from different studies looking at the diverse functions that mitochondria perform.

Mitochondrial dysfunction or dysregulation ties everything that we already know about mental and metabolic disorders together in a coherent way. *Mitochondria are the common pathway.* For the scientists or purists, a better name for this theory might be *the metabolic and mitochondrial theory of mental illness,* as this would include the myriad tasks of mitochondria and how mitochondria can be influenced by all things that affect metabolism. (I'll get to these in Part Three of the book.) However, because energy dysregulation does appear to account for the majority of symptoms of mental illness, the short and catchy phrase "brain energy theory" works for me.

How Mitochondrial Malfunction Leads to Mental Illness

I will now show you how mitochondrial dysfunction can result in all the brain changes and symptoms that we see in mental disorders. Mitochondria affect the development of the brain, the expression of different genes, synapse formation and destruction, and brain activity. They affect *structural problems* and *functional problems*. They tie together so much of what we already know and put it into one common pathway. Let's walk through some of the science.

Recall that in order to understand much of the existing neuroscience on mental disorders, I need to explain why some brain areas might be overactive, while others might be underactive, resulting in symptoms. I'm also looking for reasons why cells might develop abnormally and why some cells end up shrinking and dying, resulting in brain functions that are permanently absent. These concrete mechanisms of action will help us understand the symptoms of mental illness. They align with the framework that I outlined in Chapter Six.

You might also recall from Chapter Five that I said that human illnesses are usually due to a problem in one of three areas: the development, function, or maintenance of cells.

As it turns out, mitochondria play a role in all of these.

I will now walk you through five broad consequences of *mitochondrial dysfunction and dysregulation* that can address all of this: decreased cell maintenance, overactive brain functions, underactive brain functions, developmental problems, and cell shrinkage and cell death.

Decreased Cell Maintenance

One of the unique things about living cells, as opposed to non-living machines like cars, is that they require energy and metabolic resources to maintain themselves. Cell parts need to be repaired and replaced in constant, never-ending turnover. All of this requires energy and metabolic building blocks. One study estimates about one-third of the brain's ATP production goes to cell maintenance or "housekeeping" functions.[10] I've already described the roles of stress, cortisol, and mitochondria themselves in the process of autophagy, which is critically important in cell maintenance. But, as usual, there's more to the story.

Mitochondria interact with other organelles to facilitate routine maintenance functions. For example, they interact with lysosomes. When these interactions are prevented in experiments, waste products accumulate in the lysosomes.[11] They also interact with the endoplasmic reticulum (ER), which has many roles including the folding of proteins. Many neurodegenerative disorders are associated with misfolded proteins in the ER. When these misfolded proteins accumulate, a process called the *unfolded protein response* (UPR) tries to mitigate the damage. One group of researchers found that there is a microprotein on the outer membrane of mitochondria, PIGBOS, that plays a key role in the UPR. When this protein was eliminated, the cells were much more likely to die.[12] This strongly suggests that mitochondria play a key role in this process as well. These are just some of the ways that mitochondrial dysfunction can cause problems with cell maintenance, which can lead to all of the maintenance problems and structural defects found in people with mental disorders.

In some cases, structural defects in cells can lead to a positive feedback cycle that affects metabolism and can make it harder for cells to work. One specific example is myelin, which is an outer protective coating for neurons made by support cells called *oligodendrocytes*. Myelin makes it easier for neurons to send electrical signals. If a neuron has defects in its myelin coating, it will require more energy to work. An extreme example of this is multiple sclerosis, in which myelin is destroyed by an autoimmune process. Mitochondrial dysfunction has been associated with problems in myelin production and maintenance. Consistent with the brain energy theory, defects in myelin have been identified in the brains of people with schizophrenia, major depression, bipolar disorder, alcoholism, epilepsy, Alzheimer's disease, diabetes, and even obesity.[13]

Debris in the cell, another structural defect and maintenance issue, can impair mitochondria from moving around. For example, Alzheimer's disease is associated with the accumulation of a protein called *tau,* in addition to the beta-amyloid described earlier. Researchers who looked at the effects of tau protein on mitochondria found that it severely limited mitochondria's ability to move around a cell.[14] Their paths became obstructed from all the debris, and the tau proteins interfered with the cytoskeleton that mitochondria use for movement. When mitochondria are prevented from moving around a cell, the cell has trouble working properly, and the cell may shrink and/or die.

Overactive Brain Functions

Remember our discussion of overactivity or hyperexcitability from Chapter Six? Mitochondrial dysfunction or dysregulation can cause this! Again, this is probably the most paradoxical thing about mitochondrial dysfunction. Sometimes when mitochondria aren't working properly, parts of the brain can actually become overactive as opposed to underactive—even though they may not have enough ATP.

In the real world, hyperexcitability of cells is actually pretty common. There are many medical conditions that are reflections of hyperexcitable cells. Seizures are a clear and extreme example in the brain. A heart arrhythmia can be due to hyperexcitable heart cells. A muscle spasm is a hyperexcitable muscle cell. Chronic pain is due to a hyperexcitable nerve cell. These are all examples of cells firing when they shouldn't, or not stopping when they should.

Mitochondrial dysfunction can lead to overactivity and hyperexcitability. There are at least three ways this can happen:

1. Recall that mitochondria are involved in ion pumping and calcium regulation, which are both required to turn cells "off." If mitochondria aren't functioning properly, these processes will take longer to perform, and cells can become hyperexcitable.

2. Sometimes, overactivity or hyperexcitability can be due to dysfunction in cells that are meant to slow down other cells, such as GABA cells. If GABA cells aren't working properly, then the cells they are

supposed to inhibit will become unleashed and hyperexcitable. Recall from Chapter Six that cortical interneurons are one such example, and their dysfunction has been found in many mental and neurological disorders.

3. I discussed how maintenance problems can alter the structure of cells, such as problems with myelin or beta-amyloid. These maintenance problems can cause hyperexcitability; for example, a lack of myelin can allow ions to leak into a cell and cause it to fire when it shouldn't.

One research group directly demonstrated that mitochondrial dysfunction causes hyperexcitability by deleting a protein in mice, sirtuin 3, known to be essential to mitochondrial health. Sure enough, the mice developed mitochondrial dysfunction, hyperexcitability, and seizures, and they died early deaths.[15] Another research group turned stem cells from people with bipolar disorder and healthy controls into neurons and found that the neurons from the people with bipolar disorder had mitochondrial abnormalities and the cells were hyperexcitable.[16] Interestingly, lithium reduced this hyperexcitability.

Hyperexcitability of neurons has been found in many mental and metabolic disorders. It causes seizures and can be measured in the brains of people with epilepsy. Hyperexcitability has also been found in the brains of people with delirium, PTSD, schizophrenia, bipolar disorder, autism, obsessive-compulsive disorder, and Alzheimer's disease. It has even been measured in healthy rodents subjected to nothing more than chronic stress.[17] Hyperexcitability can be difficult to measure, but we really don't need to measure it. People can usually tell when it's occurring because something is happening in their bodies or brains that shouldn't be happening. Hyperexcitability of a pain cell causes pain. Hyperexcitability of the anxiety pathways in the brain causes anxiety. Hyperexcitability of any brain region that produces a human emotion, perception, cognition, or behavior will produce that experience.

Underactive Brain Functions

Mitochondrial dysfunction or dysregulation can slow down or reduce the function of cells. Cells require energy to work. Mitochondria provide it. They are also the on/off switch for cells. They control calcium levels and other signals. Brain cells need energy to make and release neurotransmitters and hormones and to function properly. Reduced cell function alone explains many of the altered levels of neurotransmitters and hormones seen in people with mental disorders. Additionally, mitochondria are directly involved in making some of the hormones, such as cortisol, estrogen, and testosterone, so if they are dysfunctional or dysregulated, these hormone levels may be dysregulated, too.

Developmental Problems

Beginning in the womb until early adulthood, the human brain is rapidly growing and forming connections between neurons and other brain cells. These connections are vitally important, laying the groundwork, or "hardwiring," for life. There are *developmental windows*—times when the hardwiring of the brain needs to take place in clear ways. If development doesn't occur normally, this window can close, and the brain will never again have the chance to be "normal." Mitochondria are critically important for all these tasks. As I discussed, they play critical roles in cell growth, differentiation, and synapse formation. When mitochondria are dysfunctional or dysregulated, the brain doesn't develop normally. This is important to keep in mind for neurodevelopmental disorders that begin in infancy or childhood, such as autism. Even at the end of life, our brains change in predictable ways. Cell growth and differentiation, as well as *neuroplasticity*, or the changing and adaptability of neurons, are important throughout life. When mitochondria aren't functioning properly, problems with all of this can occur. If cells or connections between cells are missing, this can result in permanently absent brain functions. These symptoms don't wax and wane, because the cells and connections needed to perform these functions simply aren't present.

Cell Shrinkage and Cell Death

Mitochondrial dysfunction can lead to cell shrinkage, or *atrophy*. If the quantity or health of mitochondria declines, the cell gets stressed. Recall that mitochondria spread themselves throughout the cell. Some are always moving around, looking for work to do. If the workforce is reduced, they can't keep the entire cell working. In some cases, mitochondria stop going to peripheral parts of the cell, such as axon terminals or dendrites. When they stop going there, those cell parts die. Inflammation ensues. Microglia, the immune cells of the brain, get to work and absorb some of these dead cell parts.[18] As more and more mitochondria become impaired, more and more of the cell shrinks. If the process continues, the cell dies.

It is well documented that the brains of people with chronic mental disorders show signs of cell shrinkage over time. Recall that they are aging prematurely. Different brain regions are affected in different people. Some areas, such as the hippocampus, are more commonly impacted, but even in people with the same diagnosis, such as schizophrenia, there can be numerous differences in the brain regions impacted.[19] This comes back to *heterogeneity*. Mitochondrial dysfunction and dysregulation explain this heterogeneity. Given that numerous factors affect mitochondrial function (which I'll get to just up ahead, in Part Three), they also affect different areas of the brain. So, depending upon the mix of risk factors or causes that people have, their brains will be affected differently. As I already mentioned, timing and development also matter. People who are affected at age fourteen will have brain differences from people not affected until thirty-nine.

Let's put this all together using our car analogy. A car has many parts—a fuel tank, fuel that goes into the tank (gasoline), an engine, a battery for electrical energy, a steering system, and a braking system for stopping. Problems can occur in different parts, and different symptoms can result. In some cases, the car may sputter or move slower (underactive function) if it gets water in the gas tank or if a spark plug has gone bad. If the battery begins to fail, it might be that the lights dim, or the windshield wipers slow down, or the radio doesn't work, or the car won't start at all (all underactive functions). These are very different symptoms, but they're all related to energy—and usually with just one cause. Now let's make the car like a living cell. Suddenly it requires energy to work *and* energy to maintain itself. Without enough energy, the tires start to deflate, the wheels get wobbly, the doors develop rust and holes (maintenance problems). The engine and battery get old. Battery acid starts to leak all over the engine (ROS). The oil hasn't been changed in a while, and engine parts are getting damaged. This lack of maintenance makes the engine even worse (a positive feedback cycle). At some point, this car will become a danger to the other cars and the flow of traffic on the highway. The brakes stop working (hyperexcitability due to impaired off switch). Eventually, it may cause a car crash and shut down the highway. The car can be towed to the junkyard for recycling (apoptosis). If someone attempts to drive it again, it will continue to pose a danger to the other cars on the highway and the flow of traffic (metabolic and mental disorders).

Brain Energy Theory Realized

Now let's see the brain energy theory in action. I'll walk you through some of the evidence focusing on three different mental disorders and how we can conceptualize what's happening to produce symptoms from a mitochondrial and metabolic point of view.

Major Depression

We know from many studies that mitochondria are not functioning properly in people with chronic depression.[20] For example, several lines of evidence have found that people with depression have lower levels of ATP, not only in brain cells, but also in muscle cells and circulating immune cells. Decreased ATP production has been found in animal models of depression as well. Autopsy studies looking at brain tissue of people with chronic depression have found specific abnormalities in mitochondrial proteins, clearly implicating mitochondrial dysfunction.[21] And as I already mentioned, levels of oxidative stress are elevated in people with depression.

Another line of evidence includes blood biomarkers of depression. Many researchers have taken blood samples from thousands of people with depression looking for abnormalities or differences compared to healthy controls. Numerous biomarkers have been identified. A meta-analysis of forty-six such studies tried to make sense of the differences to see if there might be a common pathway or theme. There was! The biomarkers related primarily to amino acid and lipid metabolism, both of which relate to mitochondrial function.[22]

One specific biomarker of great interest is acetyl-L-carnitine (ALC). This molecule is produced within mitochondria and is important to energy production. It is critical for the function of the hippocampus, a brain region often implicated in depression. One group of researchers looked at levels of ALC in depressed and non-depressed people and found that the depressed people, on average, had lower levels.[23] Furthermore, lower levels of ALC predicted the severity of depression, the chronicity of the illness, treatment resistance, and even a history of emotional neglect. A subsequent study in 460 patients with depression found that ALC levels improved with effective antidepressant treatments, and that these levels could help predict who would experience a full remission.[24] These researchers concluded, "New strategies targeting mitochondria should be explored to improve treatments of Major Depressive Disorder."

Perhaps the most direct and stunning evidence for the role of mitochondria in depression comes from an elegant study done in rats.[25] Researchers identified rats that had high levels of anxiety and depression-like behaviors and then studied a specific part of their brains, the nucleus accumbens, to see if there were differences in mitochondrial function and/or how the cells developed. They found both. The anxious/depressed rats had fewer mitochondria per cell, as well as differences in the way their mitochondria used oxygen to turn energy into ATP and in how mitochondria interacted with another organelle, the endoplasmic reticulum (ER). The neurons themselves also looked different. Following the trail further, the researchers found that the mitochondria from these rats had lower levels of mitofusin-2 (MFN2), a protein on mitochondrial membranes important to their ability to fuse with each other and with the ER. Here's the stunning part. They then injected the anxious/depressed rats with a viral vector that significantly increased the levels of MFN2. And that changed everything! The mitochondria began functioning normally, the neurons began to look normal, and the anxiety and depression-like behaviors stopped. This strongly suggests a causal role for mitochondria in depression and anxiety . . . at least in rats.

Some of the symptoms of depression fall neatly into the category of underactive functions or reduced metabolism. Changes in sleep, energy, motivation, and concentration likely all relate to reduced function of brain cells. The fatigue almost certainly extends to the muscles throughout the body, given that mitochondrial dysfunction has been found there, too. In some cases, people will describe "leaden paralysis," a situation in which they feel like their arms and legs are made of lead and it's difficult to even move them. Mitochondrial dysfunction in their muscles might explain this. If their muscles don't have enough energy, people will have difficulty moving them. *Catatonia* is an extreme version of metabolic failure—people can appear paralyzed from their illness and have severe difficulty moving or speaking.

Bipolar Disorder

Direct evidence for metabolic abnormalities in bipolar disorder (and schizophrenia) dates to 1956 when researchers noted abnormalities in lactate metabolism in patients.[26] There are many studies documenting mitochondrial dysfunction in bipolar disorder, with findings similar to those found in people with depression. However, one important question is: What makes depression different than mania? Anyone who has seen people with these two conditions knows there is a big difference.

In 2018, researchers published a review article, "A Model of the Mitochondrial Basis of Bipolar Disorder," in which they proposed that depressive states appear to be energy-deficient states, while manic states appear to involve increased energy production in the brain.[27] They cited several studies showing that manic states are associated with increased glucose and lactate utilization in the brain, both suggestive of increased mitochondrial energy production. Additionally, two neurotransmitters, glutamate and dopamine, have been found to be elevated in manic states, suggesting increased activity of these neurons. So, manic states appear to be one of the few unique situations in which mitochondria, at least in some brain cells, are producing more energy than normal. Although surprising, this is still mitochondrial dysfunction or dysregulation. The mitochondria are supposed to slow down at appropriate times, such as at night to allow for sleep. Different brain cells are supposed to stop at certain times—like the traffic in a major city. In manic states, the mitochondria are hyperactive in terms of their production of energy, and this is making cells "go" when they shouldn't. They aren't yielding or slowing down at appropriate times. Many parts of the brain appear to be overactive.

Several additional lines of evidence support this model.[28] Bipolar patients have been found to have higher than normal calcium levels, especially when they are manic—consistent with the hyperexcitability mechanism that I outlined. In fact, researchers have confirmed changes in neuronal excitability in bipolar patients. For anyone who has seen people with depression or mania, this makes perfect sense. People with mania clearly have too much energy. People with depression clearly don't have enough energy. It's fascinating that this has been found at the cellular level. In people with bipolar disorder, once the manic episode resolves, they continue to have mitochondrial dysfunction, but it results in too little energy production overall. One group of researchers recently identified a mitochondrial biomarker in blood cells showing a significant decrease in the number of mitochondria in both manic and depressed states that normalized when people were feeling well.[29] This suggests that something might be disrupting either mitochondrial biogenesis or mitophagy throughout the entire body during disease states, not just in the brain.

During manic episodes, one of the biggest dangers with too much energy is its effect on hyperexcitable cells. These are the cells that have damaged mitochondria, too few mitochondria, or structural damage due to maintenance problems. The short burst of energy during a manic phase isn't enough to correct the longstanding problems associated with mitochondrial dysfunction. It's not enough energy or time to repair the cells. Instead, it's enough to cause major problems, like psychotic symptoms, anxiety, and agitation. An easy way to think about this is to go back to our car analogy. If a car is in poor shape with deflated and misaligned tires due to poor maintenance, giving that car more gas suddenly is actually a dangerous thing. It's not ready to handle more energy or more speed. It will be more likely to crash and burn. That's what happens when you give a hyperexcitable cell too much energy.

Posttraumatic Stress Disorder

PTSD can be understood as a hyperexcitable trauma response system. This system is a normal response to life-threatening events but is now turning on when it shouldn't or failing to turn off when it should. For some people, this system appears to develop a lower threshold for firing. For example, many people with a trauma history will have clearly identified "triggers" that set off their symptoms. These can be places, people, smells, words, or even thoughts.

Two areas of the brain are commonly affected, the amygdala and the medial prefrontal cortex (mPFC). The amygdala sets off the fear response and has been found to be hyperexcitable in PTSD. The mPFC is an area of the brain that inhibits the amygdala. By doing so, it can stop a panic reaction once someone realizes that there's no need to panic. This area of the brain has been found to be underactive in people with PTSD, meaning that the person will have trouble stopping a panic reaction. Several lines of evidence have demonstrated mitochondrial dysfunction in people with PTSD, including autopsy studies showing abnormalities in mitochondrial gene expression, reductions in total mitochondria, increased levels of oxidative stress, and reduced levels of ATP.[30]

A Unifying Example

But can all mental disorders really be due to metabolism and mitochondrial dysfunction?

Some people may still have a difficult time with this new way of conceptualizing mental disorders, as lumping all the disorders together under the pathway of metabolism and mitochondrial dysfunction may seem like a stretch.

To help address this concern, we should be able to look at situations in which we know mitochondrial function is impaired abruptly and see essentially all psychiatric symptoms emerge. As it turns out, we have a clear example for this test proof—delirium.

Delirium is a serious condition defined as *an acute mental disturbance.* The word "acute" means that it happens rapidly. The "mental disturbance" can be *any* psychiatric symptom—confusion, disorientation, distractibility, fixation on certain topics, hallucinations, delusions, mood changes, anxiety, agitation, withdrawal, dramatic changes in sleep, and personality changes. Every single symptom of any psychiatric disorder can occur during delirium. Even changes in eating behaviors and perception of body image that mimic an eating disorder have been observed during delirium.

So, what causes delirium? The standard answer right now: No one knows for sure how it all works, but we know that when people are seriously ill, it can happen. Almost all medical disorders can cause delirium. This includes things like infections, cancer, autoimmune disorders, heart attacks, and strokes. The more serious they are, the more likely they are to cause delirium. People admitted to intensive care units (ICUs) are much more likely to have delirium, with anywhere from 35 to 80 percent of critically ill patients being diagnosed with it, depending upon the study.[31]

Medications can also do it. People starting new medications can have reactions that can cause delirium. Withdrawal from medications or substances, even toxic ones like heavy use of alcohol, can cause delirium. Alcohol withdrawal delirium has a special name, *delirium tremens.* It can be quite serious and life-threatening. The elderly are particularly vulnerable to delirium. People with preexisting dementia, such as Alzheimer's disease, are even more vulnerable. Essentially, there are countless causes of delirium. As I will soon explore, they all affect mitochondrial function.

How is delirium diagnosed? When symptoms of delirium begin, sometimes the cause is obvious. In some cases, the initial symptoms can be seen as a normal response to a medical condition. Many people who have heart attacks will have anxiety. It's difficult to imagine not being anxious when facing a life-threatening situation. Physicians will often give people psychiatric medications, such as benzodiazepines, to reduce their anxiety. In this early phase, they usually don't diagnose delirium or a psychiatric disorder, even though they are prescribing a psychiatric medication. The anxiety is often seen as a normal, understandable reaction. However, if these anxiety symptoms are the start of delirium, the symptoms usually become more dramatic. People can develop panic attacks and severe anxiety. This can quickly progress to confusion, disorientation, and hallucinations. This is common in frail, elderly people with heart attacks. Although these symptoms can be identical to those of dementia or schizophrenia, physicians don't assign these diagnoses. Instead, they diagnose delirium.

But how do they think about this distinction? Most healthcare professionals know that the brain simply isn't working right under the stress of a heart attack. They will attribute *any and all* new psychiatric symptoms to the heart attack. Anything goes. Obsessions. Compulsions. Confusion. Depression. Agitation. Delusions. Anything! Healthcare professionals will lump all symptoms under the diagnosis of delirium. People with delirium won't have all the symptoms of every mental disorder. They may have just a few. Some will develop symptoms of OCD. Others will look more depressed and withdrawn. Others will look manic and agitated. This doesn't matter, either. Any combination of symptoms is irrelevant. They are all due to delirium.

Delirium can sometimes occur more gradually. One of the most common causes of delirium in the elderly is a urinary tract infection, or UTI. This can be more difficult to recognize and diagnose. Oftentimes, these people don't know they have a UTI. The brain shows the first signs of a problem, not the bladder. Elderly people who were otherwise fine a few weeks prior can begin to develop confusion and memory problems. Family members or healthcare professionals often become concerned about Alzheimer's disease. The symptoms look exactly the same. These people can get confused often. They might get lost driving. They can have trouble remembering the names of people who they see every day. Once taken to a healthcare professional, a medical workup might then reveal the problem—a UTI. Treating the UTI can resolve all symptoms. Although it was due to an infection in the bladder, the symptoms came from the brain. Why? Because the brain is the most sensitive organ to energy deprivation, or mitochondrial dysfunction. It is the weakest link. It almost always shows at least some subtle signs first.

How do these different medical conditions lead to the development of essentially every psychiatric symptom? Experts speculate about neurotransmitters, stress responses, and inflammation.[32] All of these are true. But how, exactly, do they fit together and lead to psychiatric symptoms? At this point, the medical community doesn't have a coherent theory, but the brain energy theory offers one.

I am not the first to suggest that delirium is due to metabolic problems. In 1959, George Engel, the developer of the biopsychosocial model, proposed that delirium is due to disturbed brain energy metabolism, or "cerebral metabolic insufficiency."[33] Many researchers have expanded on this hypothesis since then.[34] For example, PET imaging studies have revealed decreased brain glucose metabolism in people with delirium.[35] Many serious medical conditions are known to directly impact metabolism and mitochondrial function. However, given that the medical community hasn't been able to explain mental symptoms, it has not been clear if or how these metabolic and mitochondrial abnormalities might play a role in causing the mental symptoms.

How is delirium treated? The treatment depends upon the underlying cause or causes, and the specific symptoms. Once the medical condition that is causing delirium is identified, the standard treatments for that condition are implemented. This can be an antibiotic for the UTI or standard cardiology protocols for the heart attack. What about the mental symptoms? Even though the symptoms fall under the label of delirium, we use just about anything and everything in psychiatry to control the symptoms. Sedating medications are commonly used—antipsychotics, mood stabilizers, antidepressants, anti-anxiety medications, and sleep medications. If the symptoms of delirium are extreme depression and a lack of energy, sometimes stimulants are used. We use medications to help with the symptoms, but really, we wait for the underlying medical condition to be fully treated. The symptoms of delirium usually go away once the medical condition resolves. It's a temporary case of mitochondrial dysfunction.

Does delirium really matter? Once the primary medical condition is identified, such as the heart attack, does it really matter if the person has mental symptoms or not? Many people think it doesn't. They don't take the mental symptoms very seriously. They see them as annoying problems that just make delivering good care more difficult. For example, some cardiologists will ignore mental symptoms in the context of a heart attack. To them, the problem is plain and clear—it's a heart attack. Whether the person is anxious or not doesn't really matter. Even if a person is hallucinating, they might think that doesn't matter either. That is a problem for a consulting psychiatrist to deal with. It doesn't pertain to the heart or the cardiologist's job. Unfortunately, this all-too-common perspective is shortsighted. It ignores a tremendous amount of research showing that delirium does matter. Sometimes, it's the difference between life and death.

If the brain energy theory is correct, people with delirium should have more widespread or severe mitochondrial dysfunction than people without delirium. The "mental" symptoms are giving us a warning sign. If this is true, more widespread and severe mitochondrial dysfunction should mean a whole host of things. It should mean that people with delirium will be more likely to develop a mental disorder, dementia, or a seizure. It should also mean they will be more likely to die. Are any of these things true? It turns out they all are.

Mental disorders, such as anxiety disorders, depression, and PTSD, are common after an episode of delirium. Higher rates of dementia and cognitive impairment have been consistently documented at three, twelve, and eighteen months after hospital discharge in people with delirium compared to people with the same illnesses who don't have delirium.[36] In fact, delirium in older people results in an eightfold increased risk of subsequent dementia. Hyperexcitable brain cells have also been well documented, with seizures being the most extreme consequence. In one study of people with delirium, 84 percent had abnormal electroencephalograms (EEGs), with 15 percent showing clear seizure activity.[37] People with delirium are more likely to die early. During a hospital admission, people with delirium are twice as likely to die early as those without delirium.[38] After discharge from the hospital, the one-year mortality rates for people with delirium are 35 to 40 percent, much higher than in those without delirium.[39]

How do we make sense of this? Delirium tells us that there is mitochondrial dysfunction in the brain. Sometimes it can be reversible, and the person completely recovers. But not always. What this data suggests is that the mitochondrial dysfunction can persist or progress. Mitochondria in the cells can become damaged—a reduction in the workforce of the cells. This leaves cells more vulnerable to continued dysfunction. Some of the cells themselves can actually die and not get replaced. All of this results in a reduction in the reserve capacity of different brain regions. Any of these can lead to mental disorders, Alzheimer's disease, or seizures.

What about people who show signs of a more subtle mental disorder, say depression, during an ICU stay? If depression is also due to mitochondrial dysfunction, then we should expect that depression would be associated with higher death rates or seizures. Is this true? Yes, it is. Recall that I discussed the research showing that people who are depressed after a heart attack are twice as likely to have another one over the following year. And elderly people with depression are six times more likely to have seizures. Similar research has been done in patients with a wide variety of medical illnesses. After an ICU stay, patients with depression were 47 percent more likely to die than those without depression in the two years following discharge.[40] This and other research suggests that *any* mental symptoms, even if not formally diagnosed as delirium, are associated with higher rates of premature death. Arguably, mental symptoms are like the canary in the coal mine; they are sometimes the first indication of metabolic and mitochondrial failure.

What about people with longstanding mental disorders? If their disorders are truly due to mitochondrial dysfunction, that should leave them more vulnerable to developing delirium if, in fact, delirium is due to mitochondrial dysfunction. Is this true? Yes, it is. Recall the Danish population study of more than seven million people.[41] That study found that people with mental disorders—all of them—were more likely to develop "organic" mental disorders, which includes the diagnoses of delirium and dementia. All told, depending upon the diagnosis, people were two to twenty times more likely to develop these disorders. Chronic mental illnesses are like a warning light in a car. They give us a window into a person's metabolic health. They tell us that the brain isn't working right due to metabolic or mitochondrial dysfunction. If we ignore it, sometimes it will correct itself. If it continues and we turn a blind eye to it, symptoms and other illnesses will usually follow.

If the example of delirium didn't convince you, maybe this one will—the dying process. In some medical schools, students are taught a mantra of the dying process—"seizures, coma, death." This is the sequence of events that commonly occur as people are dying. What this leaves out is delirium, which is almost universal. People will commonly hallucinate, become disoriented, have mood symptoms, or develop any of the other mental symptoms. Their brains are failing because the mitochondria in their brain cells are failing. The dying process is unequivocally associated with mitochondrial failure. This short sequence of events—delirium, seizures, coma, and death—highlights all the consequences of mitochondrial dysfunction that I have been discussing. It highlights the paradox of both reduced cell function and hyperexcitable cells in the context of rapid mitochondrial failure, culminating in death.

The Question of Language and Our Path Toward Treatment

The theory of brain energy suggests that all mental disorders have the common pathway of mitochondria. When mitochondria aren't working right, neither is the brain. If this is true, how much do our diagnostic labels matter? What should we call mental disorders?

Our current diagnostic labels will likely persist for some time. Change is difficult and takes time. Furthermore, our current diagnoses do provide some useful information. They describe constellations of symptoms that people exhibit. Symptoms matter. They will require different treatments—at least different symptomatic treatments.

However, given the overlap discussed between the diagnoses, and the fact that people with the exact same diagnoses can have different symptoms, there is clearly room for improvement. Mitochondrial dysfunction or dysregulation provides an explanation for an unlimited number of symptoms in different people. We've seen that depending upon which brain cells and brain networks are involved, and which factors are affecting the function of mitochondria, people will have different symptoms. This establishes that a change in the way we think about mental disorders is warranted.

One simple model would be to call all mental disorders delirium. Maybe we separate transient delirium and chronic delirium. The transient type resolves within two to three months, and the chronic type lasts longer. This label would remind all clinicians that they need to keep looking for the cause or causes of the metabolic brain dysfunction instead of simply administering symptomatic treatments. This would largely follow current treatment protocols for delirium already in place, but it would expand these protocols to all who are currently labeled "mentally ill."

As some will resist the use of "delirium" for all mental disorders, we could, alternatively, call all mental disorders "metabolic brain dysfunction" and add specifiers for the different symptoms that people are experiencing. For example, someone with prominent anxiety symptoms might receive the diagnosis of "metabolic brain dysfunction with anxiety symptoms." A person with schizophrenia might receive the diagnosis of "metabolic brain dysfunction with psychotic, depressive, and cognitive symptoms." In all cases, the primary diagnosis would remain the same, "metabolic brain dysfunction," but the symptoms would change as treatments work or as the illness progresses or remits. Instead of having multiple diagnoses, as is commonly the case now, people would have one disorder, metabolic brain dysfunction, with the different symptoms of this disorder.

The Brain Energy Theory . . . in a Nutshell

Here's a quick recap of the brain energy theory:

Mental disorders are metabolic disorders of the brain. Although most people think of metabolism as burning calories, it's much more than that. Metabolism affects the structure and function of all cells in the human body. Regulators of metabolism include many things, such as epigenetics, hormones, neurotransmitters, and inflammation. Mitochondria are the master regulators of metabolism, and they play a role in controlling the factors just listed. When mitochondria aren't working properly, at least some of the cells in your body or brain won't function properly.

Symptoms of mental illness can be understood as overactive, underactive, or absent brain functions. Mitochondrial dysfunction or dysregulation can cause all of these through five distinct mechanisms: (1) cell activity can be overactive; (2) cell activity can be underactive; (3) some cells can develop abnormally (leading to absent brain functions); (4) cells can shrink and die (also leading to absent brain functions); and (5) cells can have problems maintaining themselves (which can contribute to overactivity, underactivity, or absent brain functions). For example, if the cells that control anxiety are overactive, you will have anxiety symptoms. If the cells that control memory are underactive, you will have memory impairment. If metabolic problems occur at a young age, the brain can develop abnormally, which can occur in autism. If metabolic problems occur over long periods of time, cells can shrink and die, which is found in most chronic mental disorders and Alzheimer's disease. And finally, maintenance problems can leave cells in a state of disrepair and can contribute to any of these other problems.

So, you're likely wondering what causes metabolic and mitochondrial dysfunction or dysregulation. The answer is . . . lots of things. The good news is that many of them are probably already familiar to you. They will be the focus of the next and final section of the book. The exciting news is that most of them can be identified and addressed.

Part III

Causes and Solutions

Chapter 9

What's Causing the Problem and What Can We Do?

It's time to fully revisit the known risk factors and theories for what causes mental illness through the new lens of brain energy, or metabolism and mitochondria. If all mental disorders are metabolic disorders and if mitochondria are really the common pathway, then all known risk factors for mental illness must tie directly to metabolism and mitochondria, fitting together somehow. We must see evidence of cause and effect. Most of what I'm about to run through are well-established and irrefutable risk factors. Until now, though, no one has been able to connect them. In the following chapters, I will connect each one of them to metabolism and mitochondria, proving the link that has long been missing in the mental health field.

The term *risk factor* is appropriate when cause-and-effect relationships are unknown. The brain energy theory changes this. Therefore, I will start using the term *contributing causes* instead of risk factors. For most people, there are several contributing causes that come together to result in illness; it's not usually just one single root cause.

A quick note on terminology: In this section, I'll sometimes refer to metabolism and other times talk about mitochondria. They are closely related, but they are not the same thing. Going back to our traffic analogy, it might help to think of the difference this way: Metabolism is the flow of traffic, and mitochondria are the drivers and workers inside the cars. As discussed earlier, while the drivers are primarily responsible for the flow of traffic, they aren't the only factor. Traffic is also affected by environment, weather, and unforeseen obstacles, including things like daylight or nighttime driving; rain, snow, or hailstorms; road construction; and other factors that are beyond the control of the drivers but require drivers to respond as well. So, mitochondria are always involved in metabolism, but metabolic problems or challenges are not always due to mitochondrial "dysfunction." Sometimes the mitochondria are doing what they are supposed to be doing under the circumstances, but the environment poses metabolic challenges. To give an easy example, a woman could be metabolically healthy, but if she takes a hallucinogen, she might begin to hallucinate right away. The drug dysregulates her metabolism and mitochondria, causing symptoms, but it's unfair to mitochondria to say they are "dysfunctional." They are simply doing their best under the circumstances, just like drivers in cars might be doing their best in a hailstorm.

Many of the contributing causes I will discuss simply slow down mitochondria and their function. Some of them, however, are outright assaults. Some can destroy the mitochondria in cells. Some can impair their ability to produce energy. Some can impair their ability to perform other functions, such as fusing with each other or sending signals to the DNA. While some of these causes may be minor and go unnoticed initially, together with the occurrence of additional assaults they can result in enough mitochondrial impairment to produce mental symptoms. Others can be decisive and catastrophic assaults on mitochondria that result in severe mental symptoms immediately—such as mitochondrial poisons. These serious assaults often affect more than just the brain (all the cells in the body can be affected, for instance), and life-threatening situations can sometimes result.

Some of the contributing causes stimulate mitochondria and increase their production of energy, at least in the short run. This can sometimes be beneficial. It can improve symptoms of reduced cell function, such as fatigue. At other times, however, it can result in problems of too much energy. This can be as simple as the inability to sleep at night after drinking coffee. Caffeine stimulates mitochondria. But remember hyperexcitable cells? If they get too much energy, it can mean trouble—anxiety, psychosis, or seizures. As different as these symptoms are, you'll be surprised to see that sometimes one factor can trigger all of them. Prescription stimulants, such as Ritalin or Adderall, stimulate mitochondria. They can provide appropriate symptom relief in some people. However, they can also cause anxiety, psychosis, or seizures in other people.

There are three important themes to note when reviewing the different contributing causes:

1. All directly impact metabolism and mitochondria.
2. All are associated with a wide variety of symptoms of mental disorders. Not one of them is specific to any one disorder or symptom. This is consistent with the observation that *all* mental disorders have one common pathway: mitochondria.
3. All are also associated with metabolic and neurological disorders—obesity, diabetes, cardiovascular disease, Alzheimer's disease, and epilepsy. They are also associated with many other medical diagnoses, but I will focus on these five. The factors that I will be discussing can trigger exacerbations of these "physical" disorders as well. This supports the observation that mental disorders share a common pathway with these medical and neurological disorders.

I am not going to present an exhaustive scientific review of each factor. There is a tremendous amount of science to support each one. My goal here is to provide a broad overview for how all these contributing causes relate to metabolism, mitochondria, and mental health.

I will begin with the biological factors and end with psychological and social factors. This is not to imply that biological factors are more important.

In many cases, they aren't. However, reviewing the biological factors first will set the stage for how psychological and social ones affect metabolism and mitochondria.

Why Different People Have Different Symptoms and Disorders—And Why Connections Persist Across Them

I began to address this issue in Part Two, but it's worth revisiting two questions as we move into looking at specific contributing causes and treatments. First:

If all mental disorders are due to mitochondrial dysfunction or dysregulation, why is there so much variability in symptoms? How do malfunctioning mitochondria and metabolic toll cause one person to end up with depression, and another with OCD, for instance?

There are two primary answers:

1. **Differences in preexisting vulnerabilities.** All people are different. This includes identical twins. Even when the genetic code is the same for two people, they are still different. At the end of the day, we are all products of our biological blueprints (genetics) *and* our past experiences and environmental exposures. Nature *and* nurture. *Experiences* and *exposures* include the psychological and social experiences we've had, but they also include metabolic environmental exposures. These begin at conception. Our bodies are constantly responding to the environment and its access to nutrients and oxygen, hormones, temperature, light, and so many other factors. These all affect our metabolism and mitochondria, but specific factors affect some cells and not others. Over time, we develop parts of our brains and bodies that are strong and resilient, but also ones that are weaker and more vulnerable to failure. Metabolic failure of specific cells or brain networks is what causes mental symptoms,

so these areas of vulnerability influence which symptoms develop first. In essence, our metabolism is only as strong as its weakest link.

Think of it like the muscles of your body. Some of your muscles are stronger than others. If you have to lift something very heavy—a major stressor—the weakest muscles will likely fail first. This can be different in different people. If three people lift the same heavy object, one might sprain a wrist, another might pull a muscle in his leg, and a third might throw out her back. Same stressor—different symptoms—that will require different treatments due to different vulnerabilities.

2. **Differences in inputs.** Cells and the mitochondria within them are affected by many inputs, ones that impact different parts of the body and brain at different times. In the following chapters, I will discuss contributing causes that affect mitochondria in one way or another. Most of these are well-known risk factors for mental illness. Some of them will impact all the cells in the body and brain. However, most won't. Many of these factors affect only some cells but not others. When faced with different situations or tasks, different body parts or brain regions need different amounts of energy. If energy were equally distributed throughout the body and brain, not only would it be a waste of energy for the cells that don't need it, but it would divert precious energy from the cells that do need it. This means that some of these contributing causes will impact some brain regions but not others. This can lead to different symptoms.

The answers here lead directly to a second question:

If there are clear differences in individual vulnerabilities and other factors that lead to differences in dysfunction, how are all mental and metabolic disorders related? Why should having metabolic problems in one type of cell have anything to do with the function of other types of cells?

To answer this, let's go back to our analogy of metabolism: traffic in a city. There are many factors that determine whether city traffic flows smoothly or not. The same is true with metabolism. And at the same time, it's all interconnected.

A traffic problem can start in one small section of the city—a single car accident blocking a busy street. Likewise, a metabolic problem can start in one group of cells that cause symptoms related to what those cells do. The problem can start off contained to only those cells due to preexisting vulnerabilities and/or different inputs as I just outlined.

If the problems persist, however, the symptoms can spread. When it comes to city traffic, if the car accident isn't quickly removed, the traffic jam will become more widespread, affecting traffic in other parts of the city. If traffic problems are caused by poor road maintenance, then it can result in long-term problems. Metabolism is the same. A problem in one area of the body will often spread over time. Why? Because metabolism is highly interconnected. It relies on feedback loops all over the body. So, if one area is not doing well, the rest of the body can be affected. If the problem isn't corrected, it will gradually take a toll and spread—sometimes over years or decades.

Treatments and Success Stories

For each of the contributing causes of mental illness, I will outline some strategies that can be used to address problems when possible. Some of them are standard, existing treatments. (Again, a new theory doesn't replace what we already know to work.) Some of them will be new treatments that you likely haven't considered before. As a rule of thumb, they fall into the following broad categories:

1. Treatments that remove or reduce things that are dysregulating mitochondria or metabolism, such as poor diet, sleep disturbances, alcohol or drug use, some medications, or psychological/social stressors.

2. Treatments that correct for metabolic imbalances, such as neurotransmitter or hormonal imbalances.

3. Treatments that improve metabolism. I break these strategies into three categories:

    - Mitochondrial biogenesis—There are ways to increase the number of mitochondria in your cells. Increasing the workforce improves metabolic capacity.
    - Mitophagy—Getting rid of old, defective mitochondria and replacing them with new, healthy ones can also play a role. Rejuvenating the workforce improves metabolism.
    - Autophagy—Repairing structural damage that has occurred to your cells because of longstanding metabolic problems can be essential to long-term healing.

In Chapter Twenty, I'll give you an overarching approach and basic strategies to use in developing a comprehensive treatment plan. Please hold off on implementing any of the treatments that I discuss until you get to the very end of the book. You'll need to understand *all* the different contributing causes and treatment approaches before deciding which ones will be appropriate for you.

Along the way, I will also share stories of real people who have improved their mental health using metabolic interventions. Their names have been changed to protect their privacy, but their stories are true.

Chapter 10

CONTRIBUTING CAUSE

Genetics and Epigenetics

Mental disorders run in families. This has been known for centuries and established now as fact based on tremendous amounts of research. This observation has led many people to conclude that the root cause of mental illness, at least for some people, must lie in their genes. When an illness runs in families, it's largely assumed to be genetic, because genes are the way that information gets transmitted from parents to children. We now know that it's not so simple.

Genetics

From 1990 to 2003, an international community of researchers began one of the largest scientific undertakings of our time: the Human Genome Project (HGP). Researchers began to sequence and map all the human genes—three billion letters, or base pairs, of DNA. The whole world was excited and full of hope about the potential for ending all kinds of illness, particularly those thought or known to be inherited. In psychiatry, we imagined that with a complete genetic blueprint, we could identify the genes responsible for each psychiatric disorder, find out what protein they're coding for, develop medications to address the problems, and possibly even find cures.

Since the HGP was completed, researchers have discovered about 1,800 genes responsible for various diseases and developed about two thousand genetic tests that can measure an individual's genetic risk for certain diseases. Some patients can be tested to see if they will metabolize medications too slowly or fast. The effort has paid off in many ways. But sadly, not in psychiatry.

The search for the genes that cause mental illnesses has largely come up empty-handed. It's not for a lack of trying. Researchers have scoured the human genome, looking at the genes for neurotransmitters, the enzymes that make them, and their receptors. These were the obvious targets—the chemicals in the chemical imbalance theory—serotonin, dopamine, and others. Unfortunately, no meaningful relationships have been found between these genes and mental illness.

Next, researchers decided to scan the entire genome for genes that might be related to mental illness via genome-wide association studies (GWAS). They kept an open mind and looked at every gene, even ones seemingly unrelated to the brain or psychiatry, in order to find the genes that cause mental illness. After these exhaustive searches, researchers identified a plethora of genes that might be related to psychiatric disorders, but they found almost no genes that confer significant risk to a significant portion of the people with the disorders. Some very rare genes *were* identified that can confer high levels of risk, but for most people with mental disorders, specific genes don't confer large amounts of risk. On top of that, the vast majority of genes that have been found are not specific to individual disorders. Instead, they confer risk for many different psychiatric, metabolic, and neurological disorders. For example, some risk genes for psychosis confer risk for schizophrenia, bipolar disorder, autism, developmental delay, intellectual disability, and epilepsy.[1] So, one gene confers risk for many disorders. For major depression, there is debate in the field, with some studies suggesting that there are genes that confer tiny amounts of risk (I'll tell you about a couple of them soon), but other studies suggest that not even one gene has been found that confers significant risk, despite looking at more than 1.2 million genetic variations in human DNA.[2]

The disappointment of not finding genetic answers to these inheritable disorders isn't limited to psychiatry. It also applies to the metabolic disorders of obesity, diabetes, and cardiovascular disease. These, too, run in families—often the same families that have mental illness. These conditions, too, have eluded easy answers in human DNA.

Even though thousands of genes confer tiny amounts of risk, how can we understand this in the context of the brain energy theory? If mitochondria and metabolism account for all mental disorders, what do genes have to do with mental illness at all?

To begin with, many of these risk genes are directly related to mitochondria and metabolism. For example, a gene called DISC1 confers risk for schizophrenia, bipolar disorder, depression, and autism. Researchers continue to look at all the roles this protein plays in cell function, but it has been found within mitochondria and is known to influence mitochondrial movement, fusion, and contact with other parts of the cell. This, in turn, affects neuron development and plasticity.[3] One of the strongest risk genes for mood disorders, CACNA1C, plays a major role in oxidative stress and mitochondrial integrity and function.[4]

Another example is the APOE gene, which increases the risk for Alzheimer's disease. This gene codes for a protein, Apolipoprotein E, that is related to fat and cholesterol transportation and metabolism. The gene comes in three forms: APOE2, APOE3, and APOE4. About 25 percent of people carry one copy of APOE4, and 2 to 3 percent have two copies. Those with one copy are three to four times more likely to develop Alzheimer's, and those with two copies are about nine to fifteen times more likely.[5] In support of the brain energy theory, this gene strongly impacts both metabolism and mitochondria. People in their twenties who have the APOE4 gene are more likely to already show signs of impaired glucose metabolism in their brains, and this impairment worsens over time.[6]

Apolipoprotein E appears to have direct effects on mitochondria. Researchers looked at people who had different APOE types and measured proteins that impact mitochondrial biogenesis, dynamics (their fusion and fission with each other), and oxidative stress.[7] Those with the APOE4 allele had lower levels of these important mitochondrial proteins, and the levels correlated directly with symptoms of Alzheimer's. Another study looked at astrocytes, important support cells in the brain, and found that APOE4 impairs autophagy, mitochondrial function, and mitophagy.[8] The good news is that they found that giving a drug that stimulates autophagy could reverse some of these abnormalities.

So, does APOE4 increase risk for *all* of the metabolic and mental disorders? It *does* increase the risk for cardiovascular disease, *some* other mental disorders, and epilepsy, but paradoxically, it appears to *decrease* the risk for obesity and type 2 diabetes.[9] This is where the complexity of metabolism comes in. Apolipoprotein E is not found equally in all cells in the human body. In the brain, it's found primarily in astrocytes and microglia. These cells have specific functions. It appears that APOE4 results in a very slow and gradual decline in their function, and these cells are more closely related to cognitive symptoms than other symptoms. So, these "parts" of the brain are the ones that will wear down and begin to fail over time, resulting in specific symptoms. However, once they begin to fail completely, then the other mental symptoms of Alzheimer's begin due to the interconnections of these brain regions. So, as I discussed, this is an example of a preexisting vulnerability (a risk gene) and different inputs to different cells, both playing a role in causing some symptoms and not others. Nonetheless, this line of research still directly implicates metabolism and mitochondria as the common pathway to Alzheimer's.

Mitochondrial genetics are also complicated because mitochondria are influenced by genes in both the nucleus and within mitochondria themselves. The genes within mitochondria are much more susceptible to mutations. Genetic mutations in mitochondria have been directly linked to numerous aspects of brain function, including behavior, cognition, food intake, and the stress response.[10] Unfortunately, mitochondrial genes have not been widely looked at in large population studies because most researchers have assumed that mitochondrial genes don't matter all that much.

Even for genes related to other proteins supporting other functions in cells, they, too, will have an impact on metabolism. Genes are the blueprints for different proteins that make up the human body. Just like different parts of a car, and all the variations of these parts in different makes and models of cars, some are more reliable and fuel efficient than others. Some are adapted for adversity and have shorter lives while others are adapted for fuel efficiency and longevity. Depending upon which ones you inherit, they can influence the function of your cells, your metabolism, and your overall health. They can also confer different degrees of susceptibility to metabolic failure. When it comes to metabolism, there is always a "weakest link." When the parts of a cell are different, some are going to be more vulnerable to failure than others.

With this said and done, it is all but certain that for the majority of people with mental illness, the answer to their problem does not lie in the genes themselves. If genes don't fully explain why mental disorders run in families, then what could it be?

Epigenetics

Epigenetics, which we covered briefly in Part Two, is the field dedicated to understanding what causes genes to turn on or off. Most of us have similar genes. We have essentially the same blueprints for how the body should work. Yes, there are obvious variations, such as height, skin color, and hair color. These are due to differences in our genes. But most of our genes are basically the same. The human body works in the same way in most people. However, what is clearly different is the expression of all these genes.

Skin cells, brain cells, and liver cells all have the same DNA. However, epigenetics is responsible for making the different cells in the human body different from each other. These different cells express different genes.

Throughout the day, genes are getting turned on or off in cells. This is constantly changing based on environmental circumstances and the needs of the body. In other words, the body is constantly adapting. Sometimes it needs to make a hormone, so those genes get turned on. Sometimes it needs to repair a cell, so those genes get turned on. Once done, these genes get turned off. Cells don't waste resources.

There are changes in gene expression that appear to be longer lasting than these constantly fluctuating ones. Some changes in gene expression are associated with traits in people. Some people have big muscles, others are skinny, while others are obese. Gene expression is different in these people over longer periods of time, even though they all have similar underlying genes. There are specific patterns of gene expression that have been associated with different traits, both physical and mental. These longstanding epigenetic changes are a way for the body to come up with a metabolic strategy and then stick with it. Epigenetics provides a memory of what the body has been through.

There are many ways that the body controls gene expression. One way is to modify the DNA itself by applying methyl groups to specific sites on DNA. These methyl groups then influence which genes get turned on or off. The methyl groups can be added or removed as needed, but at least at some sites, they appear to become more stable over time. Another way the body influences gene expression is through histones. These are proteins that DNA gets wrapped around. They, too, influence which genes get turned on or off. In addition to methylation and histones, there are many other factors that are involved in epigenetics. More and more are being discovered every year. They include factors like micro-RNAs, hormones, neuropeptides, and others. This field quickly becomes confusing and overwhelming, as there are so many factors that are involved in the epigenetic control of our DNA.

However, if you take a step back and look at the field from a broader perspective, things become less confusing. What influences epigenetics? What triggers all these different factors to change gene expression? Almost all of them revolve around metabolism and mitochondria. Factors thought to influence epigenetics include diet, exercise, drug and alcohol use, hormones, light exposure, and sleep—all related to metabolism and mitochondria (as you will soon learn). As a specific example, smokers tend to have less DNA methylation of the AHRR gene compared to nonsmokers.[11] However, if they stop smoking, this change in methylation is reversible.

In the end, it's important to think about epigenetics as metabolic blueprints for cells. Epigenetics simply reflect the gene patterns that allow cells to do their best to survive and cope with their environments. However, if they get stuck in a maladaptive pattern or if the appropriate signals aren't being sent, that can become problematic.

Recall that mitochondria are regulators of epigenetics. They influence gene expression through levels of ROS, levels of glucose and amino acids, and levels of ATP. Also recall that mitochondria appear to control the expression of essentially all genes for a cell. I already told you about the study that found that as the number of defective mitochondria in a cell increases, the number of gene expression abnormalities also increases.

It turns out that epigenetic factors are heritable. This occurs in different ways. I'll discuss a few of them.

Womb Environment

As a fetus is growing inside the womb, it is bathed in metabolic signals. Food, oxygen, vitamins, and minerals play an obvious and critical role. However, the mother's hormones, neuropeptides, use of alcohol, drugs, prescription medications, and so many other factors are also playing roles.

One example of epigenetics playing a clear role in the transmission of both metabolic and mental disorders is the famous Dutch winter famine, which took place between 1944 and 1945 due to the German occupation of the Netherlands. Researchers have studied the babies that were conceived or carried during this famine and compared them to the general population and even their own siblings who were born when the mothers had normal access to food. The babies born during the famine were more likely to develop *both metabolic and mental disorders* later in life. This and other studies led to the *thrifty phenotype hypothesis,* which proposes that babies deprived of proper nutrition in utero are more likely to develop obesity, diabetes, and cardiovascular disease later in life. Unfortunately, this hypothesis overlooks or ignores the fact that these babies are also more likely to develop mental disorders. These babies were found to have double the risk for schizophrenia and antisocial personality disorder, as well as increased rates of depression, bipolar disorder, and addiction.[12] Researchers have been studying the pancreas to understand the elevated rates of diabetes, the heart to understand the elevated rates of cardiovascular disease, and the brain to understand the psychiatric and neurological disorders but have failed to see the *metabolic connection* among all of them.

Early Life

Some of the factors that regulate epigenetics, metabolism, and mitochondria get transferred to the baby after birth through behaviors and early life experiences. Many studies have looked at caregiver behaviors toward infants early in life and their impact on long-term health outcomes. They typically align with the ACEs studies that I already described. Caregiver neglect and deprivation have profound effects on children for life. They include both metabolic and mental disorders. Epigenetic mechanisms play a role in all of this.

A concrete example down to the molecular level is the passing of a metabolic factor from mothers to children in breast milk. One such molecule is nicotinamide adenine dinucleotide, or NAD. This is a critically important coenzyme that can be derived from vitamin B3 (niacin), or the body can make it using the amino acid tryptophan from protein. It is essential to mitochondria for energy production but also plays a role in the maintenance of DNA and epigenetics. Low levels of this enzyme are known to impair mitochondrial function and cause epigenetic changes, and have been associated with aging and many diseases.[13] One group of researchers looked at mice and the long-term outcomes of their babies after supplementing new mothers with NAD or not.[14] The mothers who got extra NAD lost more weight postpartum, a nice metabolic benefit. But their babies really benefited from this! The baby mice had improved blood glucose control, physical performance, and many brain changes, including less anxiety, improved memory, reduced signs of "learned helplessness" (a marker of depression), and they even had greater neurogenesis into adulthood. Clearly, this metabolic/mitochondrial coenzyme given in infancy impacted their brains and "mental" symptoms for life. Mothers will naturally have different levels of this coenzyme that they transfer to their babies.

Intergenerational Transmission of Trauma

The most widely studied phenomenon of how epigenetics relate to mental health is the *intergenerational transmission of trauma*. Dr. Rachel Yehuda, a leading authority in this field, outlined decades of research in a comprehensive review article: "Intergenerational Transmission of Trauma Effects: Putative Role of Epigenetic Mechanisms."[15] This field dates back to 1966, when an astute psychiatrist, Dr. Vivian Rakoff, noticed that children of Holocaust survivors appeared to sometimes have more severe forms of mental illness than their parents, even though it was the parents who spent time in the concentration camps. She asserted that somehow these things were connected. Many at the time didn't believe this. Those who did assumed that the parents must somehow be teaching their children to be afraid, anxious, or depressed, and this must be the reason for the connection. It must be a psychological or social cause. Numerous studies followed and began to identify a pattern of parental trauma and poor mental health outcomes in their children—and even their grandchildren. But still, almost everyone assumed that this was due to upbringing. The parents must be teaching their children to be stressed and afraid of the world.

This assumption was first challenged in the 1980s when researchers discovered differences in how people respond to cortisol. People exposed to trauma—*and* their children—have different levels of sensitivity to glucocorticoids. In particular, exposure to high levels of cortisol in utero appear to "program" children, resulting in higher risk for developing mental and metabolic disorders later in life. With the genetic and epigenetic revolution came the discovery that many of these people have differences in methylation patterns of the glucocorticoid receptor and other DNA regions (promoter regions) associated with the stress response system. Most recently, it has been discovered that even fathers might be passing on their traumatic experiences through epigenetic mechanisms in sperm, such as micro-RNA (miRNA) molecules, which are known to modify gene expression. Studies have now shown that sperm in both mice and men have miRNAs that get transmitted to offspring. Levels of specific miRNAs (449 and 34) have been shown to be directly affected by levels of stress dating all the way back to early childhood experiences of the fathers.[16] In mice who are exposed to early stressful life events, these levels were dramatically reduced in sperm cells, and their male offspring also had these same low levels in their sperm cells, demonstrating transgenerational transmission of stress. In human studies, men were given the ACEs questionnaire, and it turned out that men with the highest levels of stressful life events had the lowest levels of these exact same miRNAs, up to a three-hundredfold reduction.

The timing of the stress appears to matter and can influence brain function in different ways.[17] Fetal exposure to a mother's stress results in higher rates of learning impairment, depression, and anxiety later in life. In the first few years of life, separating a child from his/her mother can result in *higher* levels of cortisol throughout life, while severe abuse can result in *lower* levels of cortisol. Although paradoxical, both states take a metabolic toll and can be tied directly to mitochondria, given that mitochondria initiate production of cortisol.

This line of research continues to this day, but what it clearly demonstrates is that epigenetics appears to be playing a significant role in the transmission of mental disorders from parents to children, and even grandchildren.

What Genetics and Epigenetics Can Tell Us About Causes—and Treatment

Although some have been disappointed with our inability to find specific genes related to mental disorders, at the end of the day, I believe it's a good thing. We now know that there are usually not "abnormal" genes that cause mental illness. It's much more likely that the transmission of mental illness from parents to children takes place through epigenetic mechanisms. The hopeful aspect of this insight is that most of these epigenetic mechanisms are known to be *reversible*!

The effects of in utero stress, micro-RNA levels, and levels of NAD are changeable, sometimes through lifestyle interventions alone. The other hopeful aspect of this is that people are not usually born with "bad genes" that make it impossible for them to be healthy.

Recall my analogy of the three cars—A, B, and C. They were all the same make and model, and therefore, they all had the same blueprints (or genes). But they were very different from each other. The two primary reasons for the differences in the health, maintenance, and longevity of the cars were (1) the environment and (2) a dysfunctional driver applying adaptive strategies at the wrong times or failing to use adaptive strategies when needed. In human terms, this means that the primary causes of mental illness are usually not in our genes, but instead in our environments or the drivers of our cells, mitochondria. So, you're likely wondering what makes mitochondria dysfunctional. I'll get to many of these factors in the remainder of the book.

Even for people with the APOE4 gene allele—one that impairs mitochondrial function over time—there is hope for healing. Not everyone with this gene develops Alzheimer's disease. I mentioned the study that found that increasing autophagy can lessen the problem. I will get to more treatments in the following chapters, including some that specifically improve autophagy, but for now, understand that mental illnesses, even ones like bipolar disorder and schizophrenia, are likely not due to genetic defects that are permanent and fixed. Metabolic problems are reversible.

Chapter 11

CONTRIBUTING CAUSE

Chemical Imbalances, Neurotransmitters, and Medications

I now come back to the chemical imbalance theory. The theory of brain energy doesn't challenge the observations that neurotransmitter imbalances play a role in mental disorders, nor does it challenge the clinical trials that have demonstrated improvement in symptoms using medications that affect neurotransmitters. And I certainly don't mean to challenge the real-world experience of the many people who have been helped, or even saved, by psychiatric medications. All of these are true and based on an abundance of evidence. However, as I've already pointed out, the chemical imbalance theory leaves many questions unanswered and fails to restore lives for far too many people.

The brain energy theory offers new ways to understand neurotransmitter imbalances and the effects of medications. Mitochondria and metabolism explain the underactivity and overactivity/hyperexcitability of specific brain cells that result in these imbalances and the problem of either too much or too little neurotransmitter activity. However, neurotransmitters also go on to produce their own effects in target cells, resulting in stimulation or inhibition of mitochondria in those cells. This quickly becomes like a chain of dominoes, where metabolic disruption in one set of cells results in problems in other cells.

Many people talk about neurotransmitters as simple entities with simple functions. Serotonin makes us feel good. Dopamine drives psychosis and addiction. Norepinephrine helps us focus. Although there is some truth to these statements, these simplistic views of neurotransmitters and the disorders associated with them are almost farcical. The brain, neurotransmitters, and mental disorders are all much more complicated than that.

Neurotransmitters are not just simple on/off signals between cells. Research in the past decade has greatly expanded our view of their role in metabolism and mitochondrial function. Neurotransmitters and mitochondria are in a feedback cycle with each other. Mitochondria affect the balance of neurotransmitters. Neurotransmitters affect the balance of mitochondria and their function.

As mentioned in Chapter Seven, mitochondria play a key role in the production of many neurotransmitters, including acetylcholine, glutamate, norepinephrine, dopamine, GABA, and serotonin. Mitochondria also have receptors for some important neurotransmitters directly on their membranes, such as benzodiazepine and GABA receptors. These aren't present on all mitochondria in all cells, but they have been identified in at least some cell types. Mitochondria also have one important enzyme known to most psychiatrists: monoamine oxidase. This enzyme is involved in the degradation and regulation of some very important neurotransmitters, such as dopamine, epinephrine, and norepinephrine. All these neurotransmitters directly impact the function of mitochondria, and mitochondria directly impact the balance of these neurotransmitters.

Serotonin, a neurotransmitter best known for its role in depression and anxiety disorders, has a very prominent and complex role in metabolism and mitochondrial function.[1] It is a primitive and highly conserved neurotransmitter found in all animals, worms, insects, fungi, and plants. It is known to control appetite, digestive tract functions, and metabolism of nutrients broadly. About 90 percent of the serotonin in the human body is actually located in the digestive tract, not the brain. Recent research has demonstrated a direct role of serotonin in regulating the production and function of mitochondria within cortical neurons, enhancing their production of ATP and decreasing oxidative stress.[2] So, not only does serotonin increase mitochondrial function immediately, it results in mitochondrial biogenesis—one of the ways to improve metabolism! In addition to this clear and direct link, there's even more to the story. Serotonin is converted into melatonin, an important hormone in the regulation of sleep, which plays a powerful role in metabolism as well. Serotonin is also the product of an important metabolic pathway, the kynurenine pathway, that involves the fate of the amino acid tryptophan. When people eat protein that contains tryptophan, it has many possible fates. Two important ones are getting converted into serotonin or kynurenine. Kynurenine eventually leads to higher levels of the critically important molecule NAD that I already told you about. NAD has a profound influence on the health and function of mitochondria because it is essential to energy production and managing electrons. Problems with the kynurenine metabolic pathway have been found in many psychiatric and neurological disorders, including depression, schizophrenia, anxiety disorders, Tourette's syndrome, dementia, and others. Obviously, medications that affect serotonin levels will have a direct impact on metabolism and mitochondria through all these mechanisms. This fact likely explains why and how these medications work for disorders like depression and anxiety.

GABA is also an important neurotransmitter with a wide range of functions. It is best known for its role in anxiety disorders because medications that increase GABA activity, such as Valium, Klonopin, and Xanax, produce a calming, anti-anxiety effect. However, abnormalities of GABA neurotransmission have also been found in other disorders, including schizophrenia and autism. Mitochondria directly influence and sometimes control GABA activity. One group of researchers found that mitochondrial ROS levels regulate the strength of GABA activity.[3]

Fascinatingly, another research group demonstrated a more direct link between GABA, mitochondria, and mental symptoms. This study was done in flies and involved a known, but rare, genetic defect associated with both autism and schizophrenia. The researchers showed that mitochondria actually sequester GABA inside themselves, thereby directly controlling its release. When this process was prevented by the genetic defect, social deficits resulted. When the researchers corrected GABA levels or the function of mitochondria, the social deficits were corrected. These researchers directly tied a known, but rare, genetic defect to mitochondrial function, GABA, and social deficit symptoms.[4]

GABA doesn't just affect mental functions, it also plays a role in metabolic disorders like obesity. One group of researchers found that GABA plays an important role in brown adipose tissue, a special type of fat that gets turned on when you get cold and also plays an important role in overall body metabolism. Problems with GABA signaling in this type of fat result in mitochondrial calcium overload and metabolic abnormalities often found in people who are obese.[5] So, these few examples illustrate how mitochondria can control GABA activity, and GABA activity can affect mitochondrial function, creating a feedback cycle.

One final example is dopamine. Dopamine is released from neurons, binds to receptors, and then is usually taken back up into the releasing neuron for another round. However, some of it ends up inside cells and needs to be managed . . . by, you guessed it, mitochondria. They have the enzyme, monoamine oxidase, that degrades it. This process directly stimulates mitochondria to produce more ATP.[6] But there's more to the story. A recent discovery showed that dopamine is directly involved in the regulation of glucose and metabolism.[7] The dopamine D2 receptor is well known to most psychiatrists because almost all antipsychotic medications affect this specific receptor. We now know that dopamine D2 receptors aren't located just in the brain, they are also found in the pancreas and play a critical role in the release of insulin and glucagon. It has long been known that antipsychotic medications can affect weight, diabetes, and metabolism. The science is now catching up to explain why. More intriguing, however, is the possibility that these effects on insulin might be playing a direct role in the antipsychotic effect. It may have nothing to do with dopamine D2 receptors in the brain. I'll share more about insulin and why this might be possible in the next chapter.

These few examples illustrate some of the connections between neurotransmitters, mitochondria, and metabolism.

Psychiatric Medications, Metabolism, and Mitochondria

Medications that increase or decrease the levels of serotonin, GABA, or dopamine will clearly have an impact on mitochondria and metabolism through the mechanisms that I have already outlined. This includes many classes of antidepressants, anti-anxiety medications, and antipsychotics.

As an example, we all know that Valium can reduce anxiety. One study looked directly at the impact of Valium on anxiety and social dominance behaviors in rats to determine precisely how it was working.[8] The researchers already knew that reduced mitochondrial function in an area of the brain called the nucleus accumbens (NAc) causes social anxiety behaviors, so they wanted to determine if Valium was somehow affecting this area. They found that Valium was working by activating another area of the brain called the ventral tegmental area (VTA), which sends dopamine to the NAc. This dopamine increases mitochondrial function in the NAc, resulting in higher ATP levels, and this leads to reduced anxiety and enhanced social dominance. When the researchers blocked the effects of dopamine, this therapeutic effect was lost. But here's the kicker—when they blocked mitochondrial respiration in the NAc, the therapeutic effects were also lost, even though those cells were still getting the enhanced dopamine. These researchers concluded that their findings "highlight mitochondrial function as a potential therapeutic target for anxiety-related social dysfunctions."[9]

Different medications affect mitochondria in vastly different ways. One review article, "Effect of Neuropsychiatric Medications on Mitochondrial Function: For Better or For Worse,"[10] highlights a paradox: some medications appear to *improve* mitochondrial functions while others *impair* mitochondrial functions.

Monoamine oxidase inhibitors, a class of antidepressants, increase the amounts of epinephrine, norepinephrine, and dopamine right outside of mitochondria. These stimulate mitochondrial activity. Lithium, a mood stabilizer, has been found to increase ATP production, enhance antioxidant capacities, and improve calcium signaling within cells, all related to mitochondria.[11]

Quite a few antipsychotic medications are known to cause serious neurological problems, sometimes permanent ones, like tremors, muscle rigidity, and tardive dyskinesia (TD), an involuntary movement disorder. Many studies have documented impairment in mitochondrial function at the cellular level, including decreased energy production and increased oxidative stress, caused by these medications.[12] One study looked at spinal fluid from patients with schizophrenia, some of whom had TD, and found a direct correlation between markers of impaired mitochondrial energy metabolism and the symptoms of TD.[13] These and many other researchers have concluded that mitochondrial dysfunction is the most likely explanation for these neurological side effects.

In my own work with patients for more than twenty-five years now, I have seen firsthand how psychiatric medications can impair metabolism. Weight gain, metabolic syndrome, diabetes, cardiovascular disease, and even premature death are all well-known side effects for many of these medications.

How does this make sense? If mental symptoms are due to mitochondrial dysfunction/dysregulation, how can impairing them further reduce mental symptoms?

The answer comes down to hyperexcitable cells. When a cell is hyperexcitable, there are two ways to reduce symptoms:

1. Improve mitochondrial function and energy production so that the cell can repair itself and function normally again. However, this strategy comes with the risk that symptoms might get worse initially, given that hyperexcitable cells can't stop themselves sometimes. Therefore, when they get more energy initially, they may not be prepared to manage it appropriately, and hyperexcitability may occur.

2. Manage these cells by shutting them down—in other words, suppress their function by inhibiting their mitochondria. This will stop symptoms, at least in the short run. However, this strategy comes with the risk that it might make matters worse over time because it may worsen the mitochondrial dysfunction.

This is obviously a very concerning situation. A treatment that can help in the short run might make things worse in the long run. Unfortunately, the dilemma of hyperexcitability isn't even this simple. The brain is complicated, and so is this issue. There are two other considerations:

1. All cells are likely not impacted in the same ways. Recall that cells have different inputs. Medications target specific cells. Some cells may have improved mitochondrial function, while other cells may not be affected, while still others may have impairment. In all the studies done, the researchers had to choose specific cells to study. They certainly didn't study all the cells in the brain and body.

2. Even if medications impair mitochondrial function broadly, we need to consider the consequences of not treating the person. Hyperexcitable cells spewing out lots of neurotransmitters, such as glutamate or dopamine, are known to be toxic to the brain. The overall benefits to the person may still outweigh the risks. An extreme example is when someone is seizing—they are clearly suffering from a case of hyperexcitable brain cells. Stopping them is of paramount importance. People can die if they seize too long. And

in fact, many epilepsy treatments, such as Depakote, are known to impair mitochondrial function, which then slows, and, with hope, stops the hyperexcitability.[14]

I know that people crave simple answers to dilemmas like this. "So, should people take medications that are known to impair mitochondria? Yes or no?" Unfortunately, I can't offer a universal answer to this question because different situations require different interventions. Clearly, in dangerous or life-threatening situations, these medications can be lifesaving. However, I've mapped out some of the issues to consider. The good news is that these questions can be addressed in research studies, so more research might better inform our treatment approaches in the future. What is clear already, however, is that suppressing mitochondrial function long-term is not a path to healing. At best, it's a path to reducing symptoms.

The theory of brain energy answers numerous questions that the mental health field has been unable to answer to date. It outlines why medications that target serotonin, norepinephrine, and dopamine can all be used to treat depression. They all enhance the function of mitochondria. A logical question, then, is "Why doesn't everyone respond to the same medications?" This comes back to preexisting vulnerabilities and different inputs to different cells. For example, the symptoms of depression come from many brain regions, not just one. Brain circuits are connected and communicate with each other. If one area is malfunctioning, it will impact the other areas, too. Some will be more responsive to serotonin while others will be more responsive to norepinephrine, and yet they are connected. So, if one brain region is metabolically compromised, it will impact the other brain regions, just like a traffic jam in one part of the city slowly backing up traffic in other parts of the city. Metabolic problems are connected and can spread.

This theory also helps us understand why medications take time to work. For example, SSRIs are likely working by increasing mitochondrial biogenesis and improving the function of mitochondria. This process takes time; it doesn't occur overnight, even though SSRIs increase serotonin in a matter of hours. It's not the serotonin per se that results in improvement, but the impact that serotonin has on mitochondria and metabolism. Restoring metabolic health takes time—probably about two to six weeks—which is the time it usually takes for SSRIs to start working as well.

We can also understand why one medication can be used for a wide variety of disorders. For example, antipsychotic medications can be used for schizophrenia, bipolar disorder, depression, anxiety, insomnia, and agitation in dementia because they reduce hyperexcitability in many cell types. Suppressing mitochondrial function can stop the problematic symptoms. But anyone who has taken these medications knows that they come with side effects, like reduced function in cognitive areas of the brain and increased appetite. In the elderly, they even come with a warning of increased risk of death.

Furthermore, we can now understand why some psychiatric medications can induce other symptoms, such as antidepressants inducing anxiety, mania, and psychosis in some people. Antidepressants generally increase energy in the brain. In people with preexisting vulnerabilities who already have metabolically compromised cells, this can quickly lead to hyperexcitability and associated symptoms.

On top of the usual psychiatric medications, the brain energy theory offers reasons why "metabolic" medications might also play a role in mental health. Interestingly, psychiatrists have been using some of these for decades now.

Many blood pressure medications, such as clonidine, prazosin, and propranolol, are used in psychiatry. These medications are prescribed for a wide variety of disorders, including ADHD, PTSD, anxiety disorders, substance use disorders, and Tourette's syndrome.

One study looked at three classes of "metabolic" medications in more than 140,000 patients with schizophrenia, bipolar disorder, or other psychotic disorders to see if these medications had any impact on self-harm or the need for *psychiatric* hospitalization.[15] They found that they did. The drug classes included "statins" for cholesterol (hydroxylmethyl glutaryl coenzyme A reductase inhibitors), blood pressure medicines (L-type calcium channel antagonists), and diabetes medicines like metformin (biguanides). Across the board, these medications had an impact on the "mental" metrics, especially in reducing self-harm. The brain energy theory offers explanations for why these might help. Statins are known to impair mitochondrial function and reduce inflammation, calcium channel blockers reduce hyperexcitability by decreasing the amount of calcium in cells, and metformin is also known to play a direct role in mitochondrial function. Metformin gets confusing quickly, however, as the effects appear to be dependent upon the dose. Most studies have found that metformin impairs mitochondrial function, but some have found an increase in mitochondrial biogenesis and ATP production.[16]

Finally, I want to point out that reducing or stopping psychiatric medications can be difficult and dangerous. This always needs to be done with the supervision of a medical professional. Symptoms can get worse quickly and new symptoms can emerge. Many patients become acutely depressed, suicidal, manic, or psychotic when they stop medications abruptly or decrease them too rapidly. This doesn't mean that people can't discontinue medications; it just means that it isn't something to undertake on your own.

Summing Up

- Psychiatric medications have helped countless people with mental disorders. They will continue to play a role for many.
- The brain energy theory offers new ways to understand how and why medications work.
- It's important to understand what impact your medications are having on your metabolism and mitochondria.
- Medications that increase metabolism and improve mitochondrial function can improve symptoms of underactive cells, but they come with the risk of exacerbating symptoms related to overactive or hyperexcitable cells.

- Medications that impair mitochondrial function should be used cautiously. Although it's clear how and why these can reduce symptoms of hyperexcitable cells in the short run, it's possible that they might interfere with your ability to heal and recover in the long run. In some cases, they might even be the *cause* of symptoms. Nonetheless, in dangerous and life-threatening situations, these medications can be lifesaving.

Success Story: Jane—Agitated in the Nursing Home

Early in my career, I worked in some nursing homes as a psychiatric consultant. One person that I met was Jane, an eighty-one-year-old woman with Alzheimer's disease. I was asked to see her for "agitation." The nurses reported that she would stay up at night screaming, and at other times sleep for more than twelve hours at a stretch. Her screaming was disturbing the other residents, and they wanted me to prescribe something to stop this behavior. It had been going on for more than six months, and she had already been prescribed five tranquilizing medications, including two antipsychotics and anti-anxiety medications. Nothing was working. A medical workup had already been conducted and they couldn't find anything wrong with her.

I met with Jane for all of five minutes in the dining room, where she was propped into an adult high chair. When I sat down to speak with her, she couldn't understand me. She was saying random words and phrases (something called "word salad" in psychiatry), and she was smearing her food all over herself and her high chair. I had enough information to make my diagnosis. She was delirious. The most likely cause? The sedating medications. I wrote my note and instructed the physician to get her off as many of those medications as possible, as quickly as possible, but to be mindful that some might need to be slowly tapered. The physician ended up stopping most of them right away.

I came back three weeks later. As I was walking down the hall, I was confronted by an elderly woman I had never met before. She asked me if I was Dr. Palmer, and I said I was. She reached out and hugged me, with tears in her eyes, and thanked me profusely for saving her sister. I told her that she must be mistaken; I didn't know her or her sister. She then told me that her sister was Jane. It turns out that for the past couple of years, she had been visiting Jane three times a week. They used to have pleasant conversations and share meals, but the last six months had been a nightmare. Jane was angry, confused, and just not "human" anymore. This was heartbreaking for her sister to witness. But about ten days ago, her sister told me, things began to change. Jane had stopped screaming and her sleep was improving. She knew her sister again, and they could have conversations.

Anyone who has visited a nursing home knows this is not a rare story. It represents a common dilemma: people with dementia can become agitated and disruptive for a variety of reasons—an infection, poor sleep, or even a seemingly minor stressor, such as moving to a new room. These can all cause delirium. Jane was likely delirious when her symptoms began six months before I met her, *before* she was prescribed any psychiatric medications. Her screaming and sleep disruption were the reasons that these medications were prescribed in the first place. And they likely helped, at least temporarily. The nurses and physicians likely saw that these medications sedated Jane and decreased her screaming, and so they continued to prescribe them. When the symptoms came back, they tried increasing the doses or adding new medications.

On the surface, it's understandable how Jane ended up on so many medications. However, several of them are known to impair mitochondrial function. This means that they can help in the short run, but they come with the risk of making matters worse in the long run. That appears to be what happened to Jane. By the time I saw her, whatever caused her initial delirium had probably passed, and she was delirious because of the treatments she was receiving.

Most healthcare professionals know that tranquilizing medications can sometimes make elderly people delirious. What's going to be more difficult for the mental health field to grapple with is the possibility that this might happen in young people, too. The brain energy theory suggests that it could, and my own clinical practice for the past twenty-five years suggests that it probably does, at least in some cases. You'll hear about one such person later in the book.

This doesn't mean that antipsychotic and mood stabilizing medications shouldn't be used or can't put symptoms into remission. I believe they do work for some people, and I continue to prescribe them to this day. But for Jane, they clearly ended up making her mental symptoms even worse. Removing the offending medications brought Jane back.

Chapter 12

CONTRIBUTING CAUSE

Hormones and Metabolic Regulators

Hormones are chemical messengers that are produced in one type of cell and then travel through the body to impact other cells. The human body produces numerous hormones. All of them affect mitochondrial function and cause epigenetic changes in their target cells. Hormones change the metabolism of cells. In turn, they can play a role in both mental and metabolic disorders.

As I've discussed, mitochondria supply energy for the production and release of hormones, and they initiate the process for several key hormones.

Hormone levels are affected by a wide variety of factors. These include biological, psychological, and social factors. Hormones are one mechanism for the body to respond to stress and opportunities in the environment. In some cases, just the normal release of specific hormones can affect mood, energy, thoughts, motivations, and behaviors. Testosterone is an obvious example. Think of all the effects that it can have on men. Hormonal imbalances can be caused by many factors, including autoimmune disorders, stress, aging, and mitochondrial dysfunction in the cells that make the hormones.

There are many regulators of metabolism and mitochondrial function beyond hormones and neurotransmitters. They include things like neuropeptides, mitokines, adipokines, myokines, RNA molecules, and other messengers. Why so many factors? Because they all control different aspects of metabolic function in different cells under different circumstances. When thinking about the control of traffic, most of the stoplights in a city are activated independently of each other. However, there are some stoplights on long roads that might be coordinated with each other. These hormones and metabolic regulators, like the different stoplights, are controlling metabolism in different cells to produce desired effects. There are many roads and desired effects in the human body, hence the need for so many regulators.

I won't provide a review of all the hormones and their relationships to mental and metabolic health. That could fill an entire book. Instead, I'll briefly review a few—cortisol, insulin, estrogen, and thyroid hormone—in order to illustrate some of the connections between hormones, metabolism, and mitochondria.

Cortisol

There is no question that cortisol, metabolism, mitochondria, and mental symptoms are all interconnected, as I've discussed in previous chapters. Cortisol plays an important role in the stress response. High levels have been associated with all the metabolic disorders and numerous mental symptoms, including anxiety, fear, depression, mania, psychosis, and cognitive impairment. High levels in utero affect fetal development and play a role in epigenetics, which can lead to the later development of both metabolic and mental disorders.

Cortisol always begins in mitochondria, which have the enzyme that initiates its production. After cortisol gets released into the bloodstream, it enters cells and binds to the glucocorticoid receptor (GR), which then turns thousands of genes on or off by binding to specific sites on DNA called glucocorticoid response elements (GREs). The proteins from these genes then have widespread effects on cells, all related to metabolism. In addition to GRs in the cytoplasm and GREs located within the cell nucleus, it turns out that they are also located directly on/in mitochondria, too. In some ways, it's fair to say that cortisol begins and ends with mitochondria.

At one point in psychiatry, there was hope that cortisol would be the first definitive biomarker for mental illness. The dexamethasone suppression test, which measures fluctuations in cortisol throughout the day, was widely studied. Unfortunately, cortisol levels can actually be too high *or* too low in different psychiatric patients. Some people have high levels throughout the day while others, especially those with severe trauma histories, can have abnormally low levels. The story gets complicated quickly, and there's still debate about how and why this happens. However, my purpose is simply to illustrate that cortisol is a hormone that directly connects metabolism and mitochondria with both metabolic and mental disorders. That much is clear and unequivocal.

Insulin

Most people know insulin for its role in diabetes. People with type 1 diabetes have low levels because the pancreas is not making enough. People with type 2 diabetes are "insulin resistant," meaning that insulin isn't working effectively to allow glucose to be used as an energy source. I've already discussed the strong bidirectional relationships between diabetes and mental disorders.

Emerging evidence over the past fifteen years suggests that mitochondria are important regulators of insulin production and secretion. Mitochondria are involved in glucose metabolism and sensing how much glucose is available. They ramp up production and secretion of insulin as needed.[1]

Mitochondria are known to play a significant role in both types 1 and 2 diabetes, with some experts speculating that mitochondrial dysfunction may be the primary cause. Numerous lines of evidence support these views. One review paper outlining some of the evidence suggests that mitochondria appear to be important to the cause, complications, management, and prevention of both types 1 and 2 diabetes.[2] Insulin itself stimulates mitochondria to produce more ATP and also stimulates mitochondrial biogenesis, as measured in muscle tissue.[3] However, when researchers did this study in people with type 2 diabetes, these effects were blunted or absent. This means that diabetics, over time, may develop even more mitochondrial dysfunction due to insulin resistance—sparking a vicious cycle. It suggests that insulin resistance can be both a cause and a consequence of mitochondrial dysfunction.

But the story of insulin in brain health only begins with diabetes. It plays a powerful and direct role in brain function, too.[4] Insulin receptors are located throughout the brain, and they are involved in regulating whole-body metabolism, appetite, reproductive functions, liver functions, fat stores, and body temperature. Brain insulin also modulates neurotransmitter activity and mitochondrial function within brain cells. Changes in insulin signaling have been associated with impairment of neuronal function and synapse formation.

Insulin has been shown to influence GABA, serotonin, and dopamine neurons specifically.[5] One group of researchers demonstrated that insulin alone can increase GABA activity.[6] We know that insulin resistance can occur in the brain. When it does, it can result in mitochondrial dysfunction, which can then lead to neurotransmitter imbalances, which can then lead to overactivity and underactivity of neurons. I'll walk you through a sample of the evidence to support this.

In addition to being located on neurons, insulin receptors are also found on support cells, such as astrocytes, that play a role in providing energy for neurons. These cells can affect mood and behaviors. In animal experiments, when these insulin receptors were genetically removed, it resulted in changes in brain energy metabolism and also anxiety and depressive behaviors.[7] Insulin resistance would have similar effects.

Another animal study more directly linked brain insulin with mitochondrial dysfunction and behavioral abnormalities.[8] The researchers genetically removed brain-specific insulin receptors. This resulted in mitochondrial dysfunction, as measured by decreased ATP production and increased ROS. And sure enough, the animals displayed anxiety and depressive-like behaviors.

We have evidence that insulin resistance might be playing a role in people, too. Dr. Virginie-Anne Chouinard and colleagues of mine at Harvard and McLean Hospital did brain scans in people with schizophrenia and bipolar disorder and looked at the levels of insulin resistance in their brains.[9] They included people with recent-onset psychosis, but also their siblings who did not have psychiatric symptoms and healthy controls. We know the siblings are at higher risk of developing a mental disorder since their family members developed one. What they found was fascinating. The brains of the people with psychosis had insulin resistance compared to the healthy controls, but the normal siblings also showed signs of insulin resistance, suggesting that insulin resistance might be a risk factor that runs in families. These researchers went on to find differences in mitochondrial function between the patients with psychosis and their normal siblings. This all suggests that insulin resistance might come first, which then leads to mitochondrial dysfunction, which then leads to psychosis. Interestingly, none of the groups (patients, siblings, or controls) differed in terms of body mass index, body fat, cholesterol levels, or physical activity—so you would never be able to tell they had insulin resistance in their brains by looking at their outward appearance or talking to them about exercise.

An even more compelling study followed almost 15,000 children as they grew from the ages of one to twenty-four.[10] The researchers measured fasting insulin levels at ages nine, fifteen, eighteen, and twenty-four. They also measured the children's risk for psychosis. What they found was alarming. Children who had persistently high insulin levels (a sign of insulin resistance) beginning at age nine were five times more likely to be at risk for psychosis, meaning that they were showing at least some worrisome signs, and they were three times more likely to already be diagnosed with bipolar disorder or schizophrenia by the time they turned twenty-four. This study clearly demonstrated that insulin resistance comes first, then psychosis.

Alzheimer's disease is also known to involve insulin resistance in the brain. Some are calling it "type 3 diabetes." Strong evidence has emerged that the brains of people with Alzheimer's disease are not getting enough energy from glucose due to insulin resistance, and that this results in mitochondrial dysfunction. The areas of the brain most affected have the most plaques and tangles, the hallmarks of this disorder.[11]

Insulin as a Treatment

So, based on all this evidence, can insulin play a role in treating mental disorders?

Interestingly, the use of insulin in psychiatry isn't new. From 1927 to the 1960s, insulin coma therapy was widely used in the treatment of serious mental disorders. Clinicians would inject patients with large doses of insulin until they went into a coma. This process was repeated a few times per week. Most reports from that era suggested that it was a highly effective treatment, at least for some people. At one point, it was the most widely used treatment for psychosis and severe depression in the Western world. It fell out of favor due to the advent of psychiatric medications. By no means do I want it to come back. However, insulin *is* making a comeback in the mental health field.

Alzheimer's disease researchers have been using *intranasal insulin* in clinical trials for a few years now. Squirting insulin into the nose is the easiest and quickest way to get high levels of insulin directly into the brain, which overrides insulin resistance. Early results were promising. A pilot trial of intranasal insulin in 105 participants with mild cognitive impairment or Alzheimer's disease showed maintenance of cognitive abilities and improved brain glucose metabolism as measured by PET imaging over four months.[12] Unfortunately, a larger subsequent trial in 289 people over twelve months showed no benefit, but there were concerns that the insulin delivery device may have malfunctioned.[13]

One research study used intranasal insulin in sixty-two patients with bipolar disorder to see if it could improve their cognitive function over eight weeks. The people who got insulin showed improvement in executive function compared to those who got placebo.[14]

Clearly, more research is needed before insulin makes it into clinical practice, but some researchers are working on that.

Much more important for treatment, however, is *measuring your insulin and blood glucose* levels to identify problems like insulin resistance, hypoglycemia, and other issues. Although the correlation between what's happening in the brain and what can be measured from the blood in your veins isn't always direct, this information can be helpful, and sometimes invaluable. There are many tests and tools available—fasting glucose and insulin levels, oral glucose tolerance tests, continuous glucose monitoring devices, and others. You'll need to work with your healthcare provider to get these. If you identify a problem, there is a good chance this could be playing a role in your mental symptoms. There are many ways to address this problem that I'll discuss in the coming chapters. Lifestyle changes can be a powerful intervention—in particular, diet and exercise.

Estrogen

Most people think of estrogen as it relates to reproductive capacity in women, but that's only one of its many roles. The title of a science review article says it all, "Estrogen: A Master Regulator of Bioenergetic Systems in the Brain and Body."[15]

Estrogen has profound effects on metabolism. It is known to play an important role in mental health, obesity, diabetes, and cardiovascular disease. It also impacts brain metabolism directly and has widespread effects on mood, cognition, and other brain functions.

Mitochondria make estrogen. As with cortisol, they control the first step in its synthesis. Mitochondria also contain estrogen receptors. Like cortisol, estrogen can sometimes begin and end with mitochondria. Most estrogen receptors, however, are not on mitochondria but instead on the outside of cells. They are found widely throughout the brain on both neurons and glial cells, in both men and women. They are also found widely throughout the body. Nonetheless, many of the signaling pathways of estrogen, even when it binds to the receptor outside the cell, end up converging on mitochondria.

Estrogen levels fluctuate throughout the month in menstruating women. Many women experience "mental" and "metabolic" symptoms related to changes in estrogen levels. This can include changes in mood, appetite, and cravings. In fact, there is a diagnosis—premenstrual dysphoric disorder (PMDD)—to describe some of the mental symptoms when they are severe. But for women diagnosed with other mental disorders, symptoms can also fluctuate like clockwork around their periods. This goes for all mental symptoms—depression, anxiety, bipolar symptoms, psychotic symptoms, concentration problems, etc.—true to the theory of brain energy. As I discussed earlier, women are twice as likely to develop depression as men. These hormonal fluctuations and their effects on women's metabolism may explain some of this. Additionally, the blood loss from menstruation results in the loss of metabolic resources, such as iron, which can also take a metabolic toll.

Pregnancy and the postpartum period are times of high risk for mental symptoms, likely due to both hormonal changes and, more importantly, the metabolic toll of pregnancy. The nutrients and metabolic resources that it takes to create a child are enormous. This leaves women's bodies metabolically vulnerable. If you think about it, pregnancy comes with increased risks for both metabolic and mental disorders—weight gain (more than needed to carry a healthy child), gestational diabetes, eclampsia (which includes hypertension and seizures), and, of course, exacerbations of most mental disorders. Postpartum depression is well known, but some women experience postpartum mania or psychosis.

Menopause is associated with plummeting estrogen levels. Many women experience mental symptoms, including depression, anxiety, mania, and even psychosis. Women who had depression prior to menopause are five times more likely to become depressed around the time of menopause. Brain energy metabolism decreases broadly. One study looked at forty-three women over their transition into menopause and found that not only did brain energy metabolism decline, but this was directly correlated with a reduction in mitochondrial health.[16] After menopause, women are known to be at elevated risk for developing Alzheimer's disease compared to men. In some women, these brain metabolism abnormalities can correct themselves over time, but in others, they appear to become permanent, likely putting these women at increased risk for mental disorders and Alzheimer's disease. Researchers found a direct link between memory, estrogen, and mitochondria in rhesus monkeys.[17] They found that female monkeys with poor memory had more malformed, donut-shaped mitochondria at synapses in the prefrontal cortex. When they surgically induced menopause in the monkeys, sure enough, they showed signs of memory impairment and the numbers of malformed, donut-shaped mitochondria increased. When they gave the monkeys estrogen replacement therapy, the memory problems and the mitochondrial abnormalities both improved.

Estrogen as a Treatment

Oral contraception is used by millions of women. These pills usually contain both estrogen and progesterone. They sometimes have adverse mood effects, but ironically, are also sometimes used to treat mood symptoms, such as PMDD. So, it can be confusing: Do they help, or do they hurt? In the end, there are likely differences between women, with some experiencing benefits and others experiencing adverse effects. One study looked at more than one million women between the ages of fifteen and thirty-four taking birth control pills and found that they were somewhat more likely to experience depression or use an antidepressant compared to women not on the pills.[18] Another study looked at half a million fifteen-year-old women and found that those who were taking birth control were twice as likely to attempt suicide and three times more likely to commit suicide.[19] Birth control pills don't have the same levels of hormones that naturally occur in the body, so this may explain these findings. It's important for women with mood symptoms to work with their doctors to manage the risk of unwanted pregnancy and also their mental health needs.

Hormone replacement therapy after menopause may play a role for some women. In fact, with all this emerging evidence for the role of estrogen in the brain, doses of estrogen may need to be reevaluated to optimize brain health.

Thyroid Hormone

Thyroid hormone is known as *the* master regulator of metabolism. As far as researchers can tell, it acts on every cell in the human body. Thyroid hormone increases metabolism, revving up mitochondria. It plays a profound role in growth, development, temperature regulation, and the function of every organ, especially the brain. When people have too much or too little thyroid hormone, problems are almost always evident.

Although some of the mechanisms of action of thyroid hormone are still being worked out, what is clear and unequivocal are its effects on mitochondria. Thyroid hormone, either directly or indirectly, stimulates mitochondria to produce ATP or heat. Mitochondria have thyroid hormone receptors, so they sometimes get the signal directly. However, thyroid hormone also acts through genes in the nucleus, which then impact mitochondria. Thyroid hormone is also known to stimulate mitochondrial biogenesis, increasing the total number of mitochondria in cells.[20] It also induces mitophagy—the mitochondrial repair process.[21] As you know by now, these have powerful effects on human health.

Hypothyroidism occurs when the thyroid gland is underactive, producing less thyroid hormone than the body requires. It's most commonly due to an autoimmune disorder, but there are several other causes, too. It can lead to many metabolic and mental symptoms, including weight gain, obesity, heart disease, fatigue, brain fog, and depression. What is less well known is that it is also linked to bipolar disorder, schizophrenia, and dementia.[22] When hypothyroidism occurs during development, it can lead to profound neurological deficits (cretinism). The brain energy theory offers new ways to understand all of this. It connects all of these seemingly different illnesses through one pathway: mitochondria.

Thyroid Hormone as a Treatment

Thyroid hormone has been used as a treatment for mental disorders for decades, even when people have normal levels. It is commonly used in treatment-resistant depression and bipolar disorder. However, the field has not been able to explain how or why it works. The brain energy theory provides an obvious explanation. Not only does thyroid hormone increase metabolism immediately, but it also increases the health and number of mitochondria. When you increase the workforce, cells function better. However, increasing metabolism comes with the risk of overstimulating cells, especially hyperexcitable ones. So for some people, thyroid hormone can cause or exacerbate unwanted symptoms.

Summing Up

- Hormones and other metabolic regulators play a powerful role in metabolic and mental health.
- If you have signs or symptoms of hormonal imbalances, you should work with your healthcare professional to have those assessed and treated.
- If you have chronic mental or metabolic symptoms with no clear explanation for them, you should consider a comprehensive assessment of your hormonal status.
- It's important to take an inventory of the hormonal treatments that you are currently using, such as birth control or diabetes treatments, as they could be playing a role in your mental health (for better or worse).

Success Story: James—"It's My Thyroid"

When I first met him, James was a fifty-four-year-old man with a thirty-year history of bipolar disorder. He had recurrent depression every fall that would last until spring, despite trying more than twenty antidepressants and mood-stabilizing medications. His depression was crippling, often leaving him unable to get out of bed. He had also been diagnosed with hypothyroidism, high blood pressure, high cholesterol, and sleep apnea. His normal doses of thyroid medication—enough to increase his hormone levels into the "healthy" range—had done nothing to improve his depression, but we decided to try high-dose thyroid hormone as a treatment. It made a huge difference! His levels of thyroid hormone were now abnormally high, so we had to keep an eye out for side effects, such as heart arrhythmias and osteoporosis. But overall, he tolerated it quite well, and it changed his life. His recurrent depressions were all but gone. After about ten years of treatment with a high-dose of thyroid hormone, he was able to decrease the dose to a normal range and has continued to do well to this day. He still uses a low-dose antidepressant and sleep medication occasionally, but he hasn't experienced a severe depression in many years now. At the time I used this treatment with James, I didn't know how or why it worked. Now I do: brain energy.

Chapter 13

CONTRIBUTING CAUSE

Inflammation

Inflammation plays an important role in metabolism, mitochondrial function, mental health, and metabolic health. Therefore, it plays an important role in the theory of brain energy.

Let's start with an overarching observation—many people think of inflammation as a bad thing. Low-grade inflammation is often found in people with metabolic and mental disorders. Many speculate that neuroinflammation might be the root cause of at least some mental and neurological disorders. Cytokine storms (an overactive inflammatory response) can kill people with COVID. Lingering inflammation is one of the primary suspects for the cause of long COVID, in which people have mental and neurological symptoms for months or years after infection. Autoimmune disorders are when inflammation and the immune system are attacking a person's own body. A "leaky gut" can cause chronic inflammation. For all these reasons, we have heard that inflammation causes much of what ails us. We are told to decrease inflammation.

However, inflammation isn't always bad. It occurs all the time. It's usually a normal process that plays countless beneficial roles in the human body. It is involved in fighting off infections and healing injuries. It serves important signaling functions. It is involved in the normal stress response. Inflammatory cytokines are a way to send stress signals throughout the body and brain. Microglial cells, the brain's immune cells, play a role in brain development, learning, and memory. Without inflammation, we would die.

Inflammation, Metabolism, and Mental States

Inflammation is one way that the body allocates and uses metabolic resources, thereby directly affecting metabolism.

When inflammatory cytokines get released, more blood flows to that area of the body, bringing with it oxygen, glucose, amino acids, and fats to be used in some way. The inflammation is "calling" for these resources, and the body is allocating energy and supplies. This can occur due to infections or injuries, or in response to old or dying cells.

Inflammation can trigger the production of more immune cells and antibodies. These can be lifesaving when fighting off viruses, bacteria, and even newly formed cancer cells, but it takes energy and resources to make these. The body prioritizes dealing with these situations as they threaten the survival of the organism. Other times, the body is allocating resources to make adaptive changes, such as increasing the size of muscles after a workout or directing metabolic resources to specific brain regions for new learning. Even in these situations, inflammation is calling resources to these sites. In all these scenarios, there are fewer metabolic resources available to all the other cells in the body. In other words, inflammation takes a toll—a metabolic toll.

High levels of inflammation can change emotions, thoughts, motivations, and behaviors. For example, when people have a viral infection or cancer, the high levels of inflammation cause mental changes as well. People become lethargic, withdrawn, unmotivated, less confident, and more likely to want to climb into bed and rest. These are all adaptive. They are normal and healthy, even though they make people feel miserable. These changes allow for the conservation of metabolic resources. The body is fighting for its survival. This is no time to go out playing, exercising, or even reproducing. All available resources need to be used for survival. Some researchers have called this *conservation-withdrawal behavior* and have used these observations to better understand some of the symptoms of depression.

But it can go in the other direction, too. Mental states can cause inflammation. One fascinating study looked at humans and monkeys who were *lonely* and found that loneliness increased the stress response and induced a specific pattern of immune cell activation.[1] This left the lonely people and monkeys with chronic, low-grade inflammation. They were also more vulnerable to getting a viral infection. The researchers went so far as to infect the monkeys with a virus, and sure enough, the lonely monkeys had an impaired immune response. This helps explain why a mental symptom like loneliness has been linked to higher rates of not just mental disorders but also cardiovascular disease, Alzheimer's disease, and premature death.[2]

When inflammation occurs for a prolonged period, or when it is extreme, the metabolic toll can trigger or exacerbate mental and metabolic disorders. When infections, allergies, cancers, and autoimmune disorders flare up, there can be an increase in new-onset mental disorders or an exacerbation of mental symptoms in people with existing disorders. To give one surprising example, people with runny noses (rhinitis) from something like hay fever are 86 percent more likely to develop depression.[3] These inflammatory conditions are also well-known causes of delirium in the elderly. Likewise, there can be an increase in the symptoms of metabolic disorders. Blood sugar goes up in people with diabetes. People with cardiovascular disease can see an increase in blood pressure, experience chest pain, or suffer another heart attack.

A large population study of more than one million children in Denmark found that children hospitalized for serious infections were 84 percent more likely to develop a subsequent mental disorder and 42 percent more likely to be prescribed a psychiatric medication.[4] The biggest risk was within three months of the infection. In teens, there was an eightfold increase in OCD. In case you're thinking that all the kids were just "anxious" because they got hospitalized, the most common diagnoses included schizophrenia, OCD, personality disorders, mental retardation, autism, ADHD, oppositional defiant disorder, conduct disorder, and tic disorders. These are serious brain disorders, not just "anxiety" due to a hospitalization. And as you'll note, the diagnoses are widespread, not specific to any one disorder, which is consistent with the brain energy theory.

These are just a couple of examples of the research demonstrating that inflammatory conditions can lead to the development or exacerbation of both mental and metabolic disorders. But is there evidence that mitochondria are involved?

Inflammation and Mitochondria

Inflammation and mitochondria are in a complex feedback loop. Mitochondria are involved in many aspects of the normal inflammatory response, turning it both on and off. Inflammation, in turn, can impair mitochondrial function. Furthermore, mitochondrial dysfunction, even from other causes, can lead to inflammation. It's all a vicious cycle. I'll walk you through some of the evidence supporting this.

Mitochondria play a role in normal inflammation. In the chapter on mitochondria, I already shared with you the study demonstrating that they are responsible for the different phases of wound healing in macrophages. One scientific article, "Mitochondria in Innate Immune Responses," reviewed the many complex ways in which mitochondria are directly or indirectly involved in many aspects of the immune response, including fighting viruses and bacteria, but also playing a role in cell damage and stress.[5] Another paper published in *Cell* found that mitochondria appear to play a role in immune cell death when the time comes to turn off the immune response.[6] If mitochondria are not functioning properly in these cells, there will be problems with inflammation and immune cell function. It can lead to either an overactive or underactive immune and inflammatory response. These have been observed in many mental and metabolic disorders.

Inflammation directly impacts the function of mitochondria. For example, tumor necrosis factor (TNF), an inflammatory cytokine, has been found to directly inhibit mitochondrial function.[7] A more important example is that of interferon, another inflammatory cytokine. Its production is strongly influenced by mitochondria, but it has also been shown to directly inhibit three mitochondrial genes, resulting in changes in mitochondrial function.[8] Furthermore, interferon has been shown to directly suppress mitochondrial ATP production in some brain cells.[9] What makes this such an important example is that interferon can be given to people as a medication when treating serious infections or cancers. Shortly after starting interferon, all hell can break loose in terms of psychiatric symptoms—essentially, anything goes. The list includes depression, fatigue, irritability, insomnia, suicidal behavior, manic symptoms, anxiety, psychotic symptoms, concentration difficulties, and delirium.[10] All existing mental disorders can get worse with interferon. So here we go again. One drug, interferon, can produce every symptom known to psychiatry. Why? Mitochondria.

There are numerous other ways in which inflammation, immune cells, and cytokines can impact mitochondrial function, but for our purposes, the bottom line is that *inflammation can cause mitochondrial dysfunction.*

Inflammation can also affect brain development. In a fetus or young child, the brain can develop abnormally due to inflammatory conditions. For example, pregnant women with infections are 80 percent more likely to have a child with autism.[11] There are many animal models of autism in which researchers inject inflammatory molecules into pregnant mice in order to induce autism in the offspring. How can we tie this all together? Mitochondria.

Inflammation can also be a *consequence of mitochondrial dysfunction.* In addition to mitochondrial dysfunction in immune cells directly impacting the inflammatory and immune response, if mitochondria in other types of cells aren't functioning properly, that can also lead to the chronic, low-grade inflammation that we see in many people with metabolic and mental disorders.

Cells that are metabolically compromised can fall into a state of disrepair. They can have maintenance problems, shrink, or die. They can have high levels of oxidative stress. All of this triggers inflammation. The cells send out signals, damage-associated molecular patterns (DAMPs), that they are in need of repair. Dead cells need to be disposed of properly. Inflammation serves these purposes. In fact, mitochondria themselves, or at least parts of them, are known to be powerful DAMPs. When they are released from a struggling cell, it triggers inflammation. Inflammation in these cases is a normal response; it's not the primary problem, but instead a symptom of a metabolic problem. Interfering with it likely won't change a thing. In fact, in some cases, it might make things worse, as it can interfere with the normal healing process. The low-grade inflammation associated with all the mental and metabolic disorders is likely a consequence of widespread metabolic dysfunction. To address the problem, we need to understand what is causing the metabolic problems in the first place. This can include a wide variety of factors, like a poor diet, stress, hormonal problems, a lack of sleep, heavy use of alcohol or drugs, and other toxins. You'll hear more about some of these soon enough. In order to address the problem, we must address the metabolic dysfunction in the cells. If we can restore metabolic health, the inflammation will stop.

The Role of Inflammation in Treatment

For decades, suppressing inflammation has been an area of great interest to researchers. They have been studying antioxidants and anti-inflammatory agents for both metabolic and mental disorders. Billions of dollars have been spent on this research. The list of agents includes things like vitamin E, omega-3 fatty acids, N-acetylcysteine, and nonsteroidal anti-inflammatory medications like ibuprofen. When all is said and done, they don't appear to be all that effective as treatments. Their use in depression, schizophrenia, Alzheimer's disease, cardiovascular disease, obesity, and diabetes has been disappointing, at best, even though all these disorders have been associated with higher levels of chronic inflammation. One meta-analysis showed slight benefits for *some* of these agents for *some* mental disorders, but the improvements were minimal and usually not clinically meaningful.[12] On top of that, given that inflammation plays a role in normal brain and body functions, it's possible that suppressing it with medications may have unintended adverse consequences over the long run.

So, does inflammation matter in treatment? It does.

First, there are many lifestyle factors that I already mentioned that can cause widespread metabolic dysfunction, which can then lead to chronic inflammation. Addressing these can play a powerful role in decreasing inflammation and addressing metabolic and mental disorders. Taking an antioxidant to counteract the negative effects of these lifestyle factors simply doesn't work.

Autoimmune disorders are associated with high levels of inflammation and play a role in mental and metabolic disorders. Addressing these is important. Sometimes, it might require anti-inflammatory treatments. In other cases, it might require addressing hormone deficiencies. You'll need to work with your healthcare provider on all of this.

Chronic infections can also be a serious problem. When the body is unable to eliminate a viral or bacterial infection, it takes a metabolic toll and can lead to problems. HIV, chronic Lyme disease, hepatitis, and others can play a role in metabolic and mental health. You'll need to work with your healthcare provider to address these with optimal care.

Allergies can also lead to chronic inflammation. Sometimes, allergens can be avoided, but other times, you may need to work with your healthcare provider to select appropriate treatments.

Dental hygiene can also affect inflammation, and in turn, can impact both metabolic and mental disorders. It's important to brush and floss your teeth regularly and get regular dental checkups. This is one way to reduce a source of inflammation in your body.

Summing Up

- Inflammation plays a powerful role in mental and metabolic health.
- Inflammation always affects metabolism, and metabolic problems often increase levels of inflammation.
- For many people, poor diet, lack of exercise, poor sleep, smoking, use of alcohol or drugs, and other lifestyle factors are the primary causes of low-grade inflammation. Addressing these directly is more important than trying to reduce inflammation through other means, such as taking an antioxidant pill.
- Inflammation impacts mental states, and mental states can cause inflammation.
- Mitochondria are directly and indirectly involved with inflammation and immune cell function.
- Inflammation and mitochondria are in a complex feedback cycle, and this can play an important role in metabolic and mental health.

Chapter 14

CONTRIBUTING CAUSE

Sleep, Light, and Circadian Rhythms

Sleep, light, and circadian rhythms are all interconnected. They play a powerful role in metabolism, mitochondrial function, metabolic disorders, and mental disorders. Although the biology of these topics is complex, I'll offer a high-level overview and a sample of the evidence to demonstrate that these contributing causes all play a role in the theory of brain energy.

When we sleep at night, our bodies and brains enter a "rest and repair" state. The overall metabolic rate and temperature of the body decline while cells perform maintenance functions and make needed repairs vital to both short-term and long-term health. The brain undergoes many changes in neurons that are thought to play a role in learning and memory consolidation. Without sleep, cells can fall into a state of disrepair and begin to malfunction.

Sleep is part of the body's overall metabolic strategy. It is guided by circadian rhythms. The body has "clocks," both in the brain and in virtually all cells, that govern many biological processes. In the end, they all relate to metabolism. An area of the hypothalamus called the suprachiasmatic nucleus (SCN) plays a key role. The SCN detects light from our eyes and generates hormonal and nervous system responses. These signals, in turn, influence the peripheral clocks in all the cells of the body by turning thousands of genes throughout the body on or off. The circadian rhythm is largely driven by two things—light and food. It gets synchronized to cycles of *light or dark* and *feeding or fasting*.

An optimal amount of sleep for adults is about seven to nine hours per night but varies among people. Age, activity level, and other factors play a role. Infants and children need more sleep as their bodies are growing. The elderly require less sleep. When people are ill, they need more sleep temporarily, as sleep helps conserve energy. Sleep allows metabolic resources to be devoted to growth, maintenance, and repair functions.

When a person's safety is threatened, sleep needs to wait. Rest and repair are never as important as survival. This includes not just physical survival but also status in society. Anything that has us worried, including most psychological and social stressors, can disrupt sleep. This is normal, not a disorder.

Sleep problems can be defined as too much, too little, or poor-quality sleep. Any of them can take a metabolic toll. Problems sleeping can exacerbate all mental and metabolic disorders. Sleep deprivation can worsen depression, mania, anxiety, dementia, ADHD, schizophrenia, and substance use disorders. It can also exacerbate metabolic disorders. People with diabetes can see their blood sugars rise. People with obesity can gain more weight. People who have already had a heart attack can have another one. These are all examples of sleep problems worsening existing disorders. However, they can also be *contributing causes* to the onset of such disorders. There have been many studies of normal, healthy people being sleep deprived. If the sleep deprivation is extreme, it can result in depression, anxiety, cognitive impairment, mania, and psychosis. Genetic studies have found an association between the clock genes and autism, bipolar disorder, schizophrenia, depression, anxiety, and substance use disorders.[1] Long-term studies looking at people who don't get enough sleep have found that they are more likely to develop all the metabolic disorders as well. It can lead to and exacerbate epilepsy and Alzheimer's disease, too.

Sleep is in a feedback cycle with mental and metabolic disorders. These disorders themselves can cause sleep problems, which can then make the disorders even worse. It's well described that sleep problems are a common symptom of most mental disorders. What's less well known is that sleep problems are also more common in people with obesity, diabetes, cardiovascular disease, Alzheimer's disease, and epilepsy.

There are many different types of sleep disorders, including *obstructive sleep apnea,* when people have obstructed airways at night and stop breathing, and *restless legs syndrome,* when people can't stop moving their legs at night. The most common sleep disorder, however, is plain old insomnia.

So, we see strong bidirectional relationships between sleep, mental, and metabolic disorders. Clearly, something is going on here. We know that sleep problems result in a stress response and increase levels of inflammation. I've already discussed how those can impact mental and metabolic disorders. But there's more to the story once again. We have several lines of evidence demonstrating a feedback cycle between mitochondria, sleep, and circadian rhythms.

Sleep and Circadian Rhythms Impact Mitochondrial Function

Mitochondria are synchronized with our circadian rhythms. Energy production decreases at night to allow for sleep. It increases during the day, so that we can go out into the world to work and play.

Researchers have identified a specific protein, *DRP1,* that plays a central role in mitochondrial fission and ATP production.[2] The circadian clock controls this protein, which then synchronizes mitochondrial function with our daily rhythm. Interestingly, DRP1 is necessary for feedback to the circadian clock, suggesting that mitochondria may be influencing the clocks themselves through this feedback mechanism.

Another study looked at mice and the effects of sleep deprivation on mitochondrial function in four different brain regions. They found that sleep-deprived mice had impairment of mitochondrial function in all four brain regions, but especially in the hypothalamus, an area of the brain known to regulate metabolism and many hormones, such as cortisol.[3]

Hormones play a role in sleep and mitochondrial function as well. Abnormal levels of cortisol through the night can be caused by sleep problems. Cortisol levels can then impact brain function and cognitive impairment.[4] Melatonin, which increases at night and decreases in the morning, has been found to directly stimulate mitophagy. The lack of melatonin-induced mitophagy has been tied to cognitive deficits in mice.[5] This research suggests that poor sleep leads to mitochondrial dysfunction, which then leads to cognitive impairment, which can then lead to Alzheimer's disease. This hypothesis was further supported by another research group that deprived mice of sleep for nine months and then looked at mitochondrial function and the accumulation of beta-amyloid. Sure enough, the sleep-deprived mice had higher levels of mitochondrial dysfunction and beta-amyloid accumulation compared to the control group.[6] This research helps us understand why and how chronic sleep deprivation is a known risk factor for Alzheimer's disease.

One more example—remember NAD? This metabolic coenzyme is controlled by circadian clocks and directly influences mitochondrial activity, resulting in more ATP production.[7] So, when your circadian rhythm is off, your production of NAD will be off, which will throw off your mitochondrial function and your mental and metabolic health.

Mitochondria Play a Role in Controlling Sleep

The regulation of sleep involves numerous neurons and neurotransmitters, and many aspects are still being worked out. By no means is this a simple topic.

However, recent research does implicate at least one direct role for mitochondria. A 2019 *Nature* article looked at neurons that are known to induce sleep in the fruit fly to determine what makes them turn on and off. In other words, what makes these flies fall asleep? The researchers found that it's mitochondria. The levels of ROS in mitochondria were linked directly to specific receptors that induce sleep. The researchers summarized the importance of this finding: "Energy metabolism, oxidative stress, and sleep—three processes implicated independently in lifespan, ageing, and degenerative disease—are thus mechanistically connected."[8] What these researchers left out are mental disorders, which are also connected to all of this.

Another group of researchers looked at fruit flies with mitochondrial defects and found that they, too, had disrupted circadian rhythms and sleep patterns, further implicating mitochondria as crucial players.[9] A study in humans with mitochondrial defects found that almost 50 percent had sleep-disordered breathing problems.[10]

Light Impacts Mitochondria and the Brain

Light stimulates mitochondria, and different wavelengths of light have different effects. For example, red light tends to stimulate ATP production. However, blue light tends to inhibit ATP production and, instead, increase ROS production.[11] The different wavelengths affect different proteins on mitochondria. If too much of any spectrum of light is applied, mitochondria can produce too much ROS. This oxidative stress can damage mitochondria themselves and everything else in the cell.

The clearest example of "too much" light is in skin cells. When people lay out in the sun, the photons of light stimulate mitochondria in their skin. When exposure is excessive, it can lead to premature aging of the skin (moles and wrinkles) or even skin cancer.[12] Mitochondria are thought to play an important role in all of this.

Light exposure also affects the brain. There are at least three ways this occurs:

1. I already told you about the SCN. It detects light through our eyes and sends circadian signals throughout the brain and body. These, in turn, impact mitochondrial function.

2. Light exposure on your skin increases a molecule called urocanic acid (UCA) in the bloodstream. UCA travels to the brain where it

stimulates neurons that make glutamate. This has a direct impact on learning and memory.[13] So, light exposure can help you think better.

3. Researchers can deliver red and near-infrared light to the scalp and even inside the nose. This treatment is called *brain photobiomodulation*. These lights increase ATP production, change calcium levels, and stimulate epigenetic signals through direct actions on mitochondria. They are thought to enhance the metabolic capacity of neurons, have anti-inflammatory effects, and stimulate neuroplasticity.[14]

The Effect of Sleep, Light, and Circadian Rhythms on Symptoms

There are countless ways that modern humans mess up sleep. We take our phones to bed with us. We read in bed—with a light, of course. We wake up in the middle of the night and turn on our computers and televisions. We stay up late gaming or binge-watching Netflix. We work night shifts. We stay out all night and party. We pull all-nighters to finish important projects due the next day. We travel far distances and get jet-lagged. All these behaviors impact our circadian rhythms and sleep, taking a metabolic toll.

Others *can't* sleep, no matter how hard they try. Their minds race with worry and anxiety. They get restless. They wake up in a panic and can't get back to sleep. They snore loudly and wake up constantly. They have flashbacks of their childhood abuse. They're afraid to sleep. Their beds have become torture chambers. This also takes a metabolic toll.

On a daily basis, sleep, light, and circadian rhythms have important effects on symptoms. People with mood disorders can experience fluctuations based on the time of day; this is called *diurnal variation*. Some wake up feeling very depressed, but their moods improve as the day goes on. People with dementia can get agitated and more confused at night—something called *sundowning*. Likewise, some people with schizophrenia can get more symptomatic at night, too. The brain energy theory offers a new way to understand these well-known phenomena through mitochondria and metabolism.

Seasons can also affect symptoms. People with *seasonal affective disorder*, or depression during the winter months, are largely thought to be suffering from reduced exposure to sunlight. People with bipolar disorder can experience manic and depressive episodes around the change in seasons. The brain energy theory offers a new way to understand these changes, too.

Sleep, Light, and Circadian Rhythms as Treatment

Adequate sleep is critically important to mental and metabolic health. There are many ways it can play a role in treatment.

First, you might want to assess your sleep using these basic questions (any "no" responses are worrisome):

- Are you getting seven to nine hours of sleep every night?
- Do you sleep well through the night?
- Do you wake up feeling refreshed?
- Are you able to sleep well without the use of pills or substances?
- Do you feel reasonably awake and alert throughout the day? (Frequent naps or nodding off are worrisome signs.)

If you have chronic sleep problems, talk with your healthcare professional to determine what might be causing them. You might have obstructive sleep apnea, restless legs syndrome, a hormonal imbalance, or other reason for your sleep problem. Interventions such as *sleep hygiene* and *cognitive behavioral therapy for insomnia (CBT-I)* can play a role in treatment. These can be done in person with a therapist, but both are now accessible over the internet, too.

Sleeping pills, including over-the-counter supplements such as melatonin, can be helpful as short-term interventions for unusually stressful situations. However, sleeping pills impair the normal architecture of sleep, which may impact some of the benefits of natural sleep. They can also impair metabolism and mitochondrial function over time, so chronic use can potentially make your problems worse. Try to normalize your sleep without the use of pills. If you've been using sleeping pills for years, you may need professional help to get off them.

Next, assess your light exposure (any "no" responses may be problematic).

- Do you get exposure to natural light most days, even if just through a window?
- Do you get outside?
- Do you open the curtains or shades to allow light in?
- When you sleep, are you in a dark room with minimal or no lights?
- Do you avoid exposure to video screens while in bed (phone, television, tablet, etc.)?

Correcting any problems with light exposure, either not enough during the day or too much at night, can play a role in treatment.

Bright-light therapy is an intervention that involves sitting in front of a light every morning for about thirty minutes. These are special lights that are 10,000 lux (a measure of light intensity) that mimic exposure to sunlight but are generally safe on the eyes. Bright-light therapy has been used in a wide variety of disorders, including seasonal affective disorder, bipolar disorder, major depression, postpartum depression, insomnia, traumatic brain injury, and dementia.[15] Interestingly, light exposure might even play a role in treating obesity, diabetes, and cardiovascular disease.[16] Light therapy can help to regulate your circadian rhythms and normalize your sleep, which, as you now know, can have powerful effects on your metabolism and mitochondria. I should warn you that I have seen some patients with bipolar disorder develop hypomania and even mania from bright-light therapy, so please use it cautiously if you have had mania in the past.

I also mentioned *brain photobiomodulation*. This is still considered experimental but is being studied for a variety of conditions, such as dementia, Parkinson's disease, stroke, traumatic brain injury, and depression.

Summing Up

- Sleep, light, and circadian rhythms are all interconnected.
- They all play a powerful role in metabolism, mitochondrial function, mental health, and metabolic health.
- Diagnosing the cause of sleep problems is important, as it may require specific treatments.
- There are many things that people can do to regulate sleep.
- Controlling light exposure and/or using bright-light therapy can play a role in treatment for some people.

Success Story: Kaleb—A Twelve-Year-Old Boy Struggling in School

Kaleb lived in an upper-middle-class town and had a reasonably good life, although his parents were divorced (one on the list of adverse childhood experiences). He also had a strong family history of mental illness—his mother, father, aunts, uncles, and grandparents had suffered from depression, suicide attempts, substance abuse, bipolar disorder, and/or schizophrenia. Beginning in preschool, he had difficulties. As he grew older, he clearly met the criteria for ADHD; he ran wild at times and was often distractible. He would get frustrated with schoolwork and throw tantrums.

He started psychotherapy. His parents and teachers tried many interventions, both disciplinary strategies and behavioral rewards. Nothing worked. He started a stimulant for his ADHD, which helped for about a week, but then he couldn't sleep. That only made the problems worse. Different doses and different stimulants were tried, but the sleep problems weren't improving. Sleeping pills were considered, but his parents decided to stop the stimulant instead.

His trouble at school worsened. His IQ and learning abilities were high; they weren't the problem. He got support at school via an individualized education program (IEP), and eventually he was enrolled in a special education track for students struggling with social/emotional problems. He began reporting chronic depression. When frustrated, he would jab himself with a sharp pencil. When *really* frustrated, he would threaten suicide. In seventh grade, both the school and his therapist began recommending a mood stabilizer for presumed bipolar disorder. His parents refused and instead wanted to try a "metabolic" treatment plan.

We chose two interventions based on the metabolic basis of bipolar disorder. One issue we set out to address was insulin resistance. For the past couple of years, Kaleb had been gaining weight, especially around his waist, which is a marker of insulin resistance. He had taken to eating lots of sweets immediately after school "to deal with the stress of the day" and also after dinner "as a treat." His parents allowed it given how stressful school had been for him. To address this, we recommended that he cut out all sweets during the school week. He wasn't thrilled about this part of the treatment plan but agreed to try it. The second intervention was aimed at better regulating his circadian rhythms and sleep, both of which are known to play a role in bipolar disorder. We used bright-light therapy every morning for at least thirty minutes. This has been shown to be effective for bipolar depression, at least in some people, and it has few side effects.[17] He already played video games every morning "to wake up," so we introduced the bright light while he played his video games so that it didn't require a change in his routine.

Within one month, things began to improve. His tantrums at school stopped. His depression and his focus improved. School was becoming more manageable for him.

The following year, in eighth grade, Kaleb got his best grades ever—straight As. In 2020, two years after beginning these interventions, he started high school during the COVID-19 pandemic. Although many of his peers were struggling with depression, anxiety, and social isolation, he thrived. He got straight As again and was taken off his IEP after the first semester. The new school couldn't believe this well-behaved, top student was on an IEP in the first place.

Kaleb has been on this treatment plan for four years now and continues to thrive. Clearly, this specific plan won't work for all kids who are struggling, but it worked for Kaleb. The brain energy theory helps us understand how and why.

Chapter 15

CONTRIBUTING CAUSE

Food, Fasting, and Your Gut

What we eat, when we eat, and how much we eat have direct effects on metabolism and mitochondria. Everyone knows that diet plays a role in obesity, diabetes, and cardiovascular disease. What most people might not know is that diet also has profound effects on mental health and the brain.

This field is massive. Tens of thousands of research articles and countless textbooks have explored the effects of diet on metabolism and mitochondria. Most of this research has focused on obesity, diabetes, cardiovascular disease, Alzheimer's disease, aging, and longevity. Although these researchers don't usually see the connection with mental health, by now, I hope *you* do.

The connections go far beyond correlations. They overlap at the level of neural circuits in the brain and, of course, the entire network of metabolism and mitochondria within the human body. For example, the neural circuits that drive appetite and eating behaviors have also been directly implicated in addiction to tobacco, alcohol, and heroin.[1] That's not too surprising to most people. What might be more surprising is that the neural circuits for loneliness overlap directly with the neural circuits that warn of starvation.[2] This study, published in *Nature,* showed that chronic social isolation in the fruit fly led to increased eating *and* decreased sleep. A "social" problem led to changes in appetite and sleep. When the researchers artificially stimulated the neural circuit for social isolation, it caused the flies to eat more and sleep less. Another study identified specific GABA and serotonin neural circuits that were directly involved in obesity *and* anxiety and depression.[3] One neural circuit plays a role in how much you weigh *and* how you feel.

Some people call this field *nutritional psychiatry,* one that looks at the role of diet in mental health. Personally, I feel this is too narrow. It's more than how diet affects brain function. It's also about how our mental states affect our metabolism, which can impact appetite and feeding behavior, which can affect overall health. It's a bidirectional relationship. Metabolic affects mental, and mental affects metabolic.

As I said, this field is massive. I can't possibly do it justice in one chapter. Nonetheless, I will give you a tiny *taste* (see what I did there?) of how this field relates to mental health by running through several food-related topics and tracing how they act as contributing causes under the brain energy theory.

Vitamins and Nutrients

One of the easiest places to start is with vitamins and nutrients. Several vitamin deficiencies are known to cause mental and neurological disorders. Correcting these vitamin deficiencies can sometimes completely solve the problem. Along with hormonal imbalances, vitamin deficiencies are one of the few examples in psychiatry where there is a clearly identified problem with a simple treatment.

Three of the best-known vitamin deficiencies that can result in mental and neurological symptoms include thiamine, folate, and vitamin B12. These vitamins should be routinely checked in patients with psychiatric and neurological disorders because if they are low, there is a clear treatment for them. What do these vitamins do? They are all required for energy metabolism within mitochondria. If a person is deficient in these vitamins, they will have impaired energy production within mitochondria, aka mitochondrial dysfunction.

True to the theory of brain energy, the symptoms associated with these deficiencies are widespread and include most diagnostic categories. There are many physical symptoms as well as mental symptoms. The mental ones include depression, apathy, loss of appetite, irritability, confusion, memory impairment, sleep disturbances, fatigue, hallucinations, and delusions, to name just some of them. Deficiencies in these vitamins in pregnant women can also result in developmental abnormalities in their children, highlighting the role of mitochondria in development.

There are many other vitamins and nutrients that can easily be connected to mitochondria and metabolism, but I'll move on. As I said, this field is massive.

Food Quality

Our food supply has changed dramatically in the last fifty years. Plants have been genetically modified. Cattle, pigs, and chickens are pumped with antibiotics and growth hormones to make them fatter. Processed foods are filled with artificial ingredients and often devoid of nutrition, including fiber, vitamins, minerals, and phytonutrients. Understanding the effects of all these hormones and chemicals on human metabolism is far from clear, but research suggests they do have an impact.

Junk food is often described as "junk" not only because it lacks important nutrients, but also because it usually contains highly processed and unnatural ingredients, which have been linked to poor metabolic health. We have all heard arguments about which ingredients are bad for us. Some blame fat; others blame carbohydrates; still others blame animal-sourced products. The controversies are endless. I'll walk through three examples of dietary factors that have been directly linked to mitochondrial function and both metabolic and mental health.

Trans fatty acids (TFAs) are man-made, processed fats that were originally marketed as a healthier alternative to saturated fats. We were told that "healthy vegetable shortening" was far better than lard. For years, TFAs were ubiquitous in the US food supply. Tragically, it turns out that they are in fact toxic to human health, and they have now been banned in the US. Their use has been associated with increased risk of cardiovascular disease, depression, behavioral aggression, irritability, and Alzheimer's disease.[4] Although the exact mechanisms are still unclear, one animal study tried to figure it out by assessing what impact TFAs might have on rats and their babies.[5] Researchers gave pregnant and lactating rats either TFAs or soybean/fish oil with their diets. When the baby rats were born, they got a normal diet without TFAs. At sixty days, the babies from the mothers who got TFAs showed greater anxiety, higher levels of ROS, greater inflammation, and reduced glucocorticoid receptors in the hippocampus. This one study demonstrates how several things I've discussed so far are all interconnected. A single factor in a mother's diet ended up affecting her children's anxiety, mitochondrial function, inflammation, and glucocorticoid receptor levels, which plays a role in the stress response. Wow! Luckily, these have been banned in the US as of 2018, but could something like this account for the higher rates of depression and anxiety in the youth of America? I described how parents can transmit a vulnerability to mental illness from their own trauma history. What this research suggests is that if your mother ate trans fats while she was pregnant with you, it is possible that this might play a role in your metabolic health.

Sometimes, junk food isn't called junk because of the "bad" things in it, but instead because of the "good" things that it doesn't contain. Let's look at fiber. As you likely know, fiber is found in fruits, vegetables, and whole grains and is highly recommended these days. Most experts are certain that it plays a beneficial role in metabolic health and aging. Some studies suggest that it also plays a role in mental health. High levels of adherence to the Mediterranean diet, which includes lots of fruits, vegetables, whole grains, and olive oil, has been associated with lower rates of depression and cognitive impairment.[6] One of the biggest benefits of fiber is its conversion by microbes in the gut into butyrate, a short-chain fatty acid. Butyrate, in turn, serves as a primary fuel source for the mitochondria in gut cells (colonocytes). It also plays a role in liver cells. One research group found that butyrate directly changes mitochondrial function, efficiency, and dynamics (fusion/fission), and that these changes directly impact insulin resistance, fat accumulation in the liver, and overall metabolism.[7] There are strong connections between the gut and the brain that I'll get to shortly, but interestingly, butyrate itself appears to directly play a role in sleep! And even more fascinating is that the mechanism appears to be located in the liver or the vein going to the liver (portal vein). Researchers studied mice to work this all out.[8] They injected butyrate into their guts or their portal veins and found that the mice slept 50 to 70 percent more. When they injected butyrate into other parts of their bodies, it did nothing for sleep. Another research study found that butyrate decreases neuroinflammation in aging mice, something that might protect against Alzheimer's disease.[9]

Sometimes, it's not about one specific ingredient. Instead, it might have more to do with how much we eat. Junk food can be addictive, at least for some of us. We all know the saying, "You can't eat just one." Is overeating the problem? This can lead to high levels of insulin and blood glucose, especially in people with insulin resistance. I already told you how insulin resistance is related to mental disorders and mitochondria. Does high blood glucose play a more direct role? Some research suggests that it does.

One study in diabetic rats found that high levels of glucose directly impaired mitochondria, as measured by decreased ATP production, increased oxidative stress, and decreased antioxidant capacities, and that all of this likely damages neurons.[10]

Another study looked at human endothelial cells (cells that line arteries) to see if high glucose levels affect their mitochondrial function. They found that it does. Although it didn't change baseline energy production, when the cells were stressed, those exposed to high levels of glucose had impaired ability to produce more energy. Again, a paradox. *More* glucose, or fuel, led to *lower* levels of ATP.[11]

Another study looked at twenty people with diabetes to see what impact high blood glucose had on their mood and brain function.[12] They used a clamp technology to artificially control blood glucose and exposed all the people to both normal and high glucose levels. High glucose led to impairment in processing speed, memory, and attention, and also led to reduced energy levels, increased sadness, and anxiety. This research suggests that if people with insulin resistance overeat comfort foods, it might actually cause them to feel sad and anxious and have cognitive impairment.

Finally, a meta-analysis of forty-six studies that included over 98,000 participants who did not yet have diabetes looked at levels of blood glucose to see if this increased risk for brain changes related to Alzheimer's disease. The researchers found that higher levels of blood glucose increased risk for higher levels of amyloid and brain shrinkage.[13]

Could elevated blood glucose account for the higher rates of depression and Alzheimer's disease in people with diabetes? All of this research suggests that it might play a role.

But wait . . . it's another feedback loop! It turns out mitochondria play a direct role in controlling glucose levels. A study published in *Cell* found that mitochondria in the cells of the ventromedial nucleus of the hypothalamus (VMH), an area of the brain known to regulate glucose levels throughout the body, were critical in this regulation.[14] Their fission with each other and their levels of ROS directly control glucose levels throughout the body. So, if these mitochondria aren't functioning properly, glucose regulation will be off. This, in turn, might cause sadness, anxiety, and increase the likelihood of developing Alzheimer's disease.

Obesity

Obesity is a complex topic. Most people think of it as a problem of overeating—people are eating more calories than they are burning. But "eating more calories than they are burning" has two parts. Sometimes people *are* eating too much food—in that case the question is *why* they are eating too much. But the second part of the equation is burning calories. Some people who struggle with obesity can eat very little and still fail to lose weight. Therefore, a better question is this: *Why are those who are obese storing so much fat and/or not burning it?* The reality is that almost everyone overeats on occasion. Think of Thanksgiving. Those who are thin can feel stuffed even into the next day. This drives them to eat less. Or their metabolism increases to burn off the excess calories. Either way, they go on being thin. Obese people don't have the same responses. In fact, sometimes, when obese people lose weight, their metabolism plummets. It fights their efforts to lose weight.

These are complex topics. I'm not going to try to address them here. Instead, my goal is to highlight that obesity does play a role in metabolism and mitochondrial function, and mitochondria play a role in obesity. These things, in turn, are all related to mental health.

I've mentioned that loneliness, anxiety, depression, and sleep share some neural circuits with appetite and feeding behaviors. What would happen to someone if these neural circuits were hyperexcitable? Well, that person would feel depressed, anxious, inappropriately lonely, have trouble sleeping, and would overeat. Do you know anyone like that? I can tell you that I've met many people exactly like this in my career as a psychiatrist.

Both obesity and mental disorders are associated with mitochondrial dysfunction. When people have both, these conditions can make each other worse. Mental disorders can lead to even more weight gain. Obesity can lead to even more depression, anxiety, and bipolar symptoms. One study looked at bipolar patients, some who were obese and others who weren't, and found that the obese patients had more depressive episodes than the thin ones.[15] Obesity itself was playing a role in their mood symptoms.

One way to understand some of this is through insulin. I've already shared some information about insulin, mitochondrial function, and brain function. People with obesity usually have insulin resistance, in both their bodies and brains. One group of researchers specifically looked to see if mitochondrial dysfunction is playing a role, and sure enough, they found signs of mitochondrial impairment in both the brains and livers of insulin-resistant rats.[16]

Insulin resistance takes on a life of its own, though. The pancreas responds to insulin resistance by pumping out more insulin. If a little isn't working, then send out a lot. That helps! But the problem is that higher levels of insulin usually make insulin resistance even worse over time. They drive hunger and weight gain. And one of the problems with higher and higher levels of insulin is that insulin resistance alone inhibits mitochondrial biogenesis, compounding the metabolic problem.[17]

In the brain, insulin plays a direct role in how mitochondria respond to stress. When insulin signaling is working fine, mitochondria respond to stress effectively. One way to stress mitochondria in mice is to feed them a high-fat diet (HFD), which usually leads to obesity. Researchers studied mice with insulin resistance being fed an HFD and found impairment in the stress response in mitochondria.[18] When they gave the mice intranasal insulin, the mitochondria responded normally again, and interestingly, the mice gained less weight. So, helping mitochondria function properly helped the mice deal more effectively with the HFD.

Another research group also looked at the role of mitochondria in brain cells of mice given an HFD.[19] They found that microglial cells were causing brain inflammation in response to the HFD. This occurred *before* the mice gained any weight. When they looked further to see what was driving the changes in the microglial cells, they found that it was mitochondria. There was an increase in a specific mitochondrial protein, UCP2, that was driving changes in mitochondrial dynamics (movement, fusion, and fission). When the researchers deleted this protein, the mice no longer had brain inflammation, and shockingly, *did not develop obesity* even though they continued to have access to the high-fat diet. Instead, these mice ended up eating less and were burning more calories. This study suggests that mitochondria are, in fact, key players in how the brain and body respond to high-calorie foods. Two other studies published in *Cell* confirm a direct role of mitochondria in brain cells in the regulation of feeding behaviors, obesity, and leptin resistance.[20]

When I first read this study, I was confused. The researchers clearly demonstrated a direct role of mitochondria in both brain inflammation and subsequent obesity. However, they interfered with what the mitochondria were normally doing. So, one way to look at this study is to think that the microglial mitochondria were dysfunctional and the researchers prevented them from making a mistake. Another way to look at it is that these mitochondria were receiving erroneous signals from somewhere else in the body, maybe the gut microbiome, or gut cells, or the liver. Or maybe it was due to insulin resistance, as I described earlier. But *another* distinct possibility is that the mitochondria were doing exactly what they are programmed to do; they may have been looking out for the long-term health of the organism. At this point, we don't know what the correct response should be to a toxic diet. Maybe obesity is a better survival strategy when consuming toxic foods. We can't be certain that preventing the inflammation and obesity in this case would lead to better health outcomes or longevity, but I suppose that study would be easy enough to do. As you can see, it's complicated. And then again . . . it's not. If you want to prevent obesity in mice, don't feed them the toxic diet.

Just for the record, the "high-fat diet" can also contain other unhealthy ingredients, such as sucrose—it's often the evil combination that is so fattening. I mention this because I will soon be discussing a different type of high-fat diet that usually results in weight loss and lower levels of inflammation, so it's not the fat, per se.

Fasting, Starvation, and Eating Disorders

Fasting is going without food. For any amount of time, really. We all fast when we sleep. That's why breakfast has its name—it's breaking a fast. Fasting for longer periods of time results in many changes to metabolism and mitochondria. Interestingly, it can also have profound, beneficial effects on the human body. To most people, this is surprising. We usually think about our bodies needing food and nutrients. We've all heard that we need three square meals a day. Some of us have been told to eat six or eight times throughout the day. We need to keep fueling our bodies. We need the energy.

For infants, this is unequivocally true—most need feedings every two hours. For children who are growing, this can be true as well. But for adults, there is now a tremendous amount of science to suggest that eating all the time actually harms health.

Fasting prompts the body to be frugal and encourages autophagy, which has tremendous healing potential. The body hunkers down and makes do with what it has. It's time to tap those fat stores. We all know that can be good. But there's more to the story than that. Each and every cell responds, and mitochondria are right there to direct things. They immediately change shape. They elongate themselves, fuse with each other, and form long tubular networks.[21] What ensues is a spring-cleaning, if you will. The cells identify old and defective proteins and cell parts. They are the first to go in a massive recycling campaign. These proteins and parts are shuttled to lysosomes for degradation. These nutrients are then recycled; some get used for energy, while others might be used for new, essential proteins and cell parts. The cells are looking for anything and everything that is expendable in a carefully orchestrated reboot.

So, what about the cells' mitochondria? Are they also getting destroyed? Well, the defective ones are, as mitophagy also gets activated during this process. But the healthy mitochondria are now working with each other in the long tubular networks. They keep up the pace of ATP production during this process, and these networks protect them from being recycled. When the person eats again, the cell parts that got destroyed get replaced. These replacement parts are new and healthy, and they often include some new mitochondria!

However, if people go without food for too long, at some point, it turns into starvation. The body mounts a defense strategy. It lowers metabolism broadly to conserve energy. Heart rate slows. Body temperature declines. People become sluggish, irritable, unmotivated, distractible, obsessed with food, and somewhat depressed. Paradoxically, symptoms of hypomania can emerge within the first week or two of starvation. This is likely an adaptive strategy to give the starving person enough energy, motivation, and confidence to get food, no matter what.

As I'm sure you know, this is not good. It's life-threatening. Cells begin to malfunction and die off. Organs are struggling, and this includes the brain. The list of mental symptoms and diagnoses includes depression, irritability, insomnia, mania, eating disorders, confusion, memory disturbances, hallucinations, and delusions.

Perhaps the best evidence for the mental effects of starvation come from the famous Minnesota Starvation Experiment, in which thirty-six healthy men were subjected to semi-starvation diets (half their normal daily calories) for twenty-four weeks, and then received twenty weeks of "rehabilitation." The men lost significant amounts of weight and showed signs of slowed metabolism. They experienced wide-ranging and sometimes severe mental symptoms, including depression, anxiety, fatigue, poor concentration, and obsessions with food. Some became transiently hypomanic. Interestingly, it was during the refeeding period that some had the most difficulty. Depression got worse in some of them. Others began binging and purging. Some developed body image concerns. One man cut off three fingers. This study is often used today to understand some of the symptoms of anorexia and bulimia. Starvation itself can cause mental symptoms.[22]

This leads to a discussion of eating disorders.

For some people, an eating disorder can start through societal pressure to lose weight and be thin. Young, female ballet dancers are an obvious example. Many young women who dance are told in no uncertain terms that they must be thin to compete. There is tremendous pressure to lose weight. For the girls and women who follow these recommendations, they can put themselves into starvation. This can start the vicious cycle of metabolic disturbances that impact brain function. The parts of the brain that control feeding behaviors are impacted, but so are the parts of the brain that interpret how they perceive their bodies. They can develop severe distortions in body image, thinking they are fat when they are truly emaciated. This sometimes verges on delusional, because the way they perceive themselves can be so far removed from how they really look. Researchers looked at a mouse model of anorexia to see if, in fact, there is mitochondrial impairment in the brain, and sure enough, they found oxidative stress and impairment of specific parts of mitochondria within the hypothalamus.[23] A study of forty women, half with anorexia and the other half without, found mitochondrial dysfunction in the white blood cells of those with anorexia.[24]

However, other people can develop an eating disorder because of a preexisting vulnerability. Eating behaviors impact metabolism, whether it's overeating or undereating. For some, this can result in short-term, rewarding feelings.

Binge eating can make some people feel better because it provides more insulin and glucose to struggling brain cells, and it stimulates the reward centers in the brain. This may be the quickest and easiest way to overcome insulin resistance—eat a lot of sugar. Unfortunately, as I just discussed, this makes matter worse over time. For others, restricting eating can improve mood because it can provide stress hormones or ketones (you'll hear more about these soon) that can be helpful to struggling brain cells. These two extremes of overeating and undereating can both produce rewarding experiences in different people. So, for people with preexisting mental disorders who are already struggling, changing eating behaviors can be appealing. For some, it can become a way of life, even if it causes health problems. This likely explains why people with *all* mental disorders are more likely to develop an eating disorder. They are looking for ways to feel better.

The Gut-Brain Axis and the Microbiome

Over the past few decades, a growing body of research suggests that our intestinal tract plays an important role in both metabolic and mental health. Many signals get sent from the digestive tract to the brain, and vice versa. There appear to be many mechanisms by which this communication occurs. I'll briefly look at a few of them.

First, it has become increasingly clear that the trillions of bacteria, fungi, and viruses in our guts play an important role in human health, especially obesity, diabetes, and cardiovascular disease. For example, animal studies have shown that gut microbes can affect weight. In one study, researchers found that the microbiome in obese mice extracts more nutrients and calories from food than in thin mice. When this obese microbiome is transferred to thin mice, the thin mice gain weight.[25]

There is also increasing evidence for the role of the gut microbiome in mental disorders. Animal models and small human trials have shown that the gut microbiome appears to play a role in depression, anxiety, autism, schizophrenia, bipolar disorder, and eating disorders. There is also evidence for the role of the gut microbiome in epilepsy and neurodegenerative disorders.

Gut bacteria get first dibs on all the food we eat. They produce a variety of metabolites, neurotransmitters, and hormones that they secrete into our guts. These get absorbed into our bloodstream and can affect our metabolism and brain function.

A second way that signals get sent from the gut to the brain is through hormones and neuropeptides that are produced in the cells that line the intestinal tract. These, too, are known to travel throughout the body, including to the brain, and have widespread effects on metabolism and brain function.

Finally, the gut has an intricate nervous system unto itself that communicates directly with the brain and vice versa. The vagus nerve plays an important role in this communication. As I mentioned, about 90 percent of the body's total serotonin is produced in the intestinal tract.

This field of the gut-brain axis and the microbiome can quickly become overwhelming when one begins to think about all the different microbes, metabolites, hormones, neurotransmitters, neuropeptides, and other factors involved. However, there is a clear connection between all these factors. They all relate to metabolism and mitochondria. There is evidence that gut microbes send signals directly to mitochondria in the cells lining the gut and immune cells. These signals have been shown to change mitochondrial metabolism, alter the barrier function of the gut cells, and can lead to inflammation.[26]

Food, Fasting, and the Gut Microbiome as Treatments

There are at least eight different ways that dietary interventions can be helpful in addressing mental symptoms:[27]

1. Addressing nutritional deficiencies, such as folate, vitamin B12, and thiamine deficiency.
2. Removing dietary allergens or toxins. For example, some people have an autoimmune disorder called celiac disease that results in inflammation and other metabolic problems in response to gluten. This can also affect brain function. I've described the toxic effects of TFAs. There are many other dietary ingredients that can also impair mitochondrial function.
3. Eating a "healthy diet," such as the Mediterranean diet, may play a role for some people.
4. Exploring fasting, intermittent fasting (IF), and fasting-mimicking diets (more below on all three) that stimulate autophagy and mitophagy to improve metabolic health.
5. Improving the gut microbiome (more on how below).
6. Improving metabolism and mitochondrial function with a dietary intervention. This includes changes in insulin resistance, metabolic rate, the number of mitochondria in cells, the overall health of mitochondria, hormones, inflammation, and many other known regulators of metabolism.
7. Dietary changes that result in losing weight can help to mitigate the problems associated with obesity.
8. Dietary changes that result in gaining weight can be a life-saving intervention for those who are severely underweight.

A comprehensive discussion of all these areas is beyond the scope of this book. Instead, I'll discuss a few highlights.

Vitamins and Nutraceuticals

Addressing vitamin and nutritional deficiencies is important. However, taking twenty-plus vitamins and supplements is not the answer to most metabolic problems. Sometimes, excessive use of vitamins and supplements can actually cause metabolic problems. Healthy metabolism is about balance—not too much and not too little.

Many vitamins and supplements, or *nutraceuticals, might* play a role in improving mitochondrial function and production. The list of possibilities is long. It includes L-methylfolate, Vitamin B12, SAMe, N-acetylcysteine (NAC), L-tryptophan, zinc, magnesium, omega-3 fatty acids, nicotinamide riboside, alpha-lipoic acid, arginine, carnitine, citrulline, choline, co-enzyme Q10, creatine, folinic acid, niacin, riboflavin, thiamine, resveratrol, pterostilbene, and antioxidants.[28] It's unlikely that all of these will be beneficial for all people, and by no means should anyone take all of these at the same time.

Here's a great example to illustrate this caution. Researchers gave 180 patients with bipolar depression one of three treatments: (1) a "mitochondrial cocktail," (2) NAC alone, or (3) placebo for sixteen weeks as an add-on to their existing treatments.[29] The mitochondrial cocktail included *N*-acetylcysteine, acetyl-L-carnitine, Co Q10, magnesium, calcium, vitamin D3, vitamin E, alpha-lipoic acid, vitamin A, biotin, thiamine, riboflavin, nicotinamide, calcium pantothenate, pyridoxine hydrochloride, folic acid, and vitamin B12. Wow! That's a cocktail. And guess what they found? No difference in any of the groups.

Again, low levels of these vitamins and other factors may simply be a consequence of mitochondrial dysfunction, not the cause of it. If that's the case, adding more may not solve the problem. And pills like this don't automatically stimulate mitochondrial biogenesis or mitophagy. But dietary interventions, good sleep, stress reduction, removing mitochondrially impairing medications, and exercise can!

Diet and Fasting

I shared with you that people who adhere to the Mediterranean diet (MD) are less likely to develop depression. But for people who are already depressed, can adopting this diet improve symptoms? It appears that it can, at least for some people. One trial, cleverly called the SMILES trial, randomized sixty-seven people with major depression to groups that encouraged people to follow the MD or a social support group (the control group). Participants continued their existing treatments for depression (medications or therapy). After twelve weeks, 32 percent of the people in the MD group achieved remission compared with only 8 percent in the control group.[30] Is there any evidence that this is due to metabolism or mitochondria? Well, we have at least one study to call upon.

Researchers looked at monkeys (cynomolgus macaques) fed an MD versus a Western diet (standard American diet) for thirty months and then measured brain mitochondrial function, energy utilization patterns, and biomarkers such as insulin levels.[31] They found diminished bioenergetic patterns between brain regions in the monkeys getting the Western diet, which correlated with levels of insulin and glucose. The mitochondria in the MD-fed animals maintained normal differences between brain regions, whereas the mitochondria in the Western-fed animals lost these normal distinctions. The brain areas affected are known to play a role in diabetes and Alzheimer's disease.

There is also evidence that fasting, intermittent fasting (IF), and fasting-mimicking diets may play a role in treating mental disorders. They all result in the production of ketone bodies, which are made when fat is being used as an energy source. Fat gets turned into ketones. And, interestingly, this process occurs exclusively in mitochondria, yet another role for these magnificent organelles.

Ketones are an alternate source of energy to cells. They also serve as important metabolic signaling molecules, resulting in epigenetic changes. Ketones can be a rescue energy source to insulin-resistant brain cells. While glucose might have trouble getting into these cells, ketones can get inside pretty easily. Fasting also results in autophagy, as I've described. There are several versions of IF. Some restrict eating to eight to twelve hours per day. Others allow one meal a day. And others restrict nighttime eating.

We have evidence that IF improves mood and cognition, and protects neurons from damage in animal models of epilepsy and Alzheimer's disease. One group of researchers set out to understand how and why.[32] You'll never guess what they found—it's mitochondria! When researchers put mice on an IF routine, they found that the hippocampus, a brain region often involved in depression, anxiety, and memory disorders, was largely driving the improvements from IF. It appeared to be due primarily to higher levels of GABA activity, which reduced hyperexcitability. Then the researchers went further to understand what was causing this change in GABA activity. They removed sirtuin 3 from the mice in two different ways—recall that this protein is exclusive and essential to mitochondrial health. When they did this, all the benefits were lost. This clearly implicates mitochondria directly in the benefits of IF on brain health.

Another review article outlines many of the ways that IF is thought to promote brain health, including reducing oxidative stress and inflammation, improving mitophagy and mitochondrial biogenesis, increasing brain-derived neurotropic factor (BDNF), improving neuroplasticity, and promoting cellular stress resistance.[33] These are powerful healing interventions not currently available in a pill.

Fasting-mimicking diets can replicate the benefits of fasting for longer periods of time without the risk of starvation. The best-known example is the *ketogenic diet*. You likely recall that it was this diet and its profound impact on one of my patients that led me on this journey.

The story of the ketogenic diet begins with epilepsy. Since the time of Hippocrates, fasting was known to stop seizures and was used as a treatment in many cultures. With the advent of modern medicine, however, this was largely believed to be religious folklore and likely nonsense. That changed in the 1920s when a physician published a research article about fasting stopping seizures in a boy. The problem with fasting is that if it's done too long, people die of starvation—not a very effective intervention. And when people start eating normally again, the seizures usually come right back. In 1921, Dr. Russell Wilder developed the ketogenic diet, a diet high in fat, moderate in protein, and low in carbohydrates, to address this challenge. His desire was to see if the diet could mimic the fasting state, but prevent starvation, to treat epilepsy. Lo and behold, it worked. The ketogenic diet reduced or stopped seizures in about 85 percent of the people who tried it. By the 1950s, it had fallen out of favor as a growing number of antiepileptic medications came to market. Taking a pill is much easier than doing this diet.

Unfortunately, about 30 percent of patients with epilepsy don't improve with any of the pills we have to offer, so the ketogenic diet was resurrected in the 1970s at Johns Hopkins for use in treatment-resistant epilepsy. Clinical use of this diet has grown around the world since. Many clinical trials have shown efficacy, and a 2020 Cochrane Review (a gold-standard meta-analysis) concludes that children with treatment-resistant epilepsy who eat ketogenic diets are three times more likely to achieve seizure freedom and six times more likely to experience a 50 percent or greater reduction in seizures than children getting usual care.[34]

The ketogenic diet is now the best-studied dietary intervention for its effects on the brain. Neurologists, neuroscientists, and pharmaceutical companies have been studying this diet for decades trying to better understand its anticonvulsant effects. It provides an alternate fuel source, which can be a lifeline to insulin-resistant brain cells. It also changes neurotransmitter levels, regulates calcium channels, decreases inflammation, improves the gut microbiome, increases overall metabolic rate, reduces insulin resistance itself, and most importantly, induces both mitophagy and mitochondrial biogenesis.[35] After people are on this diet for months or years, their cells have more healthy mitochondria. This can result in long-term healing. Many people can stop the diet after two to five years and remain well.

The research for the diet's efficacy in treating mental health disorders is in the early stages. In my own work, I have seen people with severe, treatment-resistant psychotic disorders achieve full remission of their symptoms for long periods of time through a ketogenic diet.[36] You'll hear about one at the end of this chapter. The effects of the diet in the first year are just like a medication. People need to remain on the diet religiously. They can't stop it for "cheat days," just like they can't stop their medications for cheat days. All hell can break loose if they do. I should point out that it's common to use epilepsy treatments in psychiatry. We use many of them for almost all types of mental disorders. So, in many ways, this is nothing new. It just happens to be a dietary intervention. There are several clinical trials for bipolar disorder and schizophrenia getting underway currently.

Alzheimer's disease researchers looked at twenty-six patients who all got twelve weeks of the ketogenic diet and twelve weeks of a low-fat diet, separated by a ten-week washout period.[37] Participants did the diets in different orders and the assessments were blinded. At the end of the study, the researchers found that when people were on the ketogenic diet, they had improvement in daily function and quality of life. I should point out that this is one of the few studies to demonstrate *improvement* in symptoms of Alzheimer's disease. Most studies, like the intranasal insulin studies I told you about, only prevent progression of the disease. They don't reverse it. Obviously, this was a small trial that needs to be replicated in more people over longer periods of time, but the basic science certainly supports why and how this might be working.

There are many versions of the ketogenic diet, including ones for weight loss, diabetes management, and epilepsy, and they don't always have the same effects. The foods can also be tailored to personal preferences, such as vegetarian, vegan, meat-only (the "carnivore diet"), or a diet with both animal and plant-sourced foods. Those with medical or mental disorders should only do this diet with medical supervision, as there are risks and side effects, and prescription medications usually need to be adjusted or stopped safely.

The Gut Microbiome

As mentioned, there is no question that the gut microbiome plays a role in mental and metabolic health. In terms of proven interventions, however, this field is in its infancy.

Here are four types of interventions to consider:

1. Avoid antibiotic exposure if possible. Antibiotics are known to disrupt the microbiome, and they can sometimes directly cause mitochondrial dysfunction. In addition to not taking antibiotics unless necessary, it's important to avoid eating foods that contain antibiotics, such as meat, fish, eggs, milk, and other products that commonly contain antibiotics fed to the animals. Look for "raised without antibiotics" labels.

2. Diet plays a critical role in the microbiome. Avoid highly processed foods. Eating foods high in fiber, such as fruits and vegetables, and a diet of real, whole foods is likely optimal.

3. Probiotics might play a role for some people, although we don't have much evidence that they can improve metabolism or mental health. Recall that there are trillions of microbes in your gut. Taking a supplement with just one type of bacteria may or may not help. Before starting one, look into any research on that specific probiotic to see if there is evidence for its effectiveness, especially for your symptoms or diagnosis.

4. Fecal microbial transplants are being studied but are experimental at this point.

Summing Up

- Diet plays a powerful role in metabolism and mitochondrial health.
- If you have any dietary deficiencies, they need to be identified and corrected. This can include vitamins, minerals, protein, or essential fatty acids, to name just a few. You may want to work with a dietitian or your healthcare provider to fully assess your nutritional status and diet.

- If you are exposing yourself to dietary factors that are harmful to your metabolism, you need to remove these from your diet. This can include allergens, but also foods that are known to be toxic, such as TFAs and junk food.
- If you have insulin resistance, you likely need to change your diet to help address the underlying problem.
- Even if you are following a perfectly healthy diet, your metabolism and mitochondria can become impaired. This can be due to non-dietary factors such as genetics, epigenetics, inflammation, stress, sleep problems, hormones, medications, toxins, etc. Even in cases like these, dietary interventions can still play a role in treatment. For example, IF and the ketogenic diet can both stimulate autophagy and mitophagy, regardless of what caused the problem in the first place. They can also provide ketones as a rescue fuel source to insulin-resistant cells.
- Strategies to improve gut health may improve mental health.
- Be skeptical of probiotics or "mito" supplements that claim to fix all your problems with one pill. In most studies to date, they don't work.
- Mental health and metabolic health are inseparable. This applies to everyone, including those simply trying to lose weight, or manage their diabetes, or prevent a heart attack or Alzheimer's disease. Diet and exercise are often not enough. Everything that I am discussing in this book plays a role.

Success Story: Mildred—It's Never Too Late!

Mildred had a horrible, abusive childhood. There is no doubt that she suffered from symptoms of PTSD, anxiety, and depression. At age seventeen, she was also diagnosed with schizophrenia. She began having daily hallucinations and delusions. She became chronically paranoid. Over the ensuing decades, she tried different antipsychotic and mood-stabilizing medications, but her symptoms continued. She could no longer care for herself and was assigned a court-appointed guardian. She was miserable. She tried to kill herself numerous times, once drinking a bottle of cleaning fluid. On top of her mental symptoms, she was obese, weighing 330 pounds.

At age seventy, after fifty-three years of being tormented and disabled by her schizophrenia, her doctor encouraged her to go to a weight-loss clinic at Duke University. They were using the ketogenic diet as a weight-loss method. She decided to give it a try. Within two weeks, not only did she begin to lose weight, she noticed significant improvement in her psychotic symptoms. She said that for the first time in years, she was able to hear the birds singing outside. The voices in her head were no longer drowning them out. Her mood was also improving, and she began to have hope. She was able to taper off all her psychiatric medications. Her symptoms went into *full remission*. She also lost 150 pounds and has kept it off to this day.

Now, thirteen years later, she remains symptom-free, off medication, and doesn't see any mental health professionals. Having learned to take care of herself, she got rid of the guardian, too. When I last spoke with Mildred, she said she was happy and excited to be alive. She asked me to share her story with anyone and everyone willing to listen. She hopes that her story might help others escape the living hell that she had to endure for decades.

Stories like Mildred's . . . just don't happen in psychiatry. Even with the best traditional treatments that we have to offer, this is unheard of. Mildred's story and the theory of brain energy say it is possible. It is a new day in the mental health field, one filled with hope for more stories like Mildred's.

Chapter 16

CONTRIBUTING CAUSE

Drugs and Alcohol

It's well known that drugs and alcohol can lead to mental disorders, and that people with mental disorders are more likely to use drugs and alcohol. Think of the young man smoking too much marijuana who ends up with schizophrenia. Or the alcoholic who develops dementia. Or the cocaine addict with bipolar disorder. Most people think these are just the results of toxic drugs on the brain. Or that maybe these people were predisposed to mental illness and the drugs pushed them over the edge. Both assertions are true. But exactly how do they cause mental illness? Up until now, no one could say for sure. The brain energy theory offers clear answers: drugs and alcohol converge on metabolism and mitochondria.

Most drugs fit into one of two categories—they either stimulate or inhibit cells. This includes alcohol, tobacco, marijuana, cocaine, amphetamines, and opiates. Some drugs work on specific cells in the brain or body, while others have broader effects on different types of cells. For example, alcohol and marijuana, which I'll focus on soon, both have broad effects throughout the body. They act through receptors mostly found on the surfaces of cells, which then impact the mitochondria within those cells. However, mitochondria also have their own receptors for marijuana, nicotine, alcohol, and Valium right on their membranes. These drugs directly impact mitochondria.

Drugs and alcohol form a feedback loop with metabolism and mitochondria. People can enter this feedback loop in different ways, but once they get in, it can be difficult to get out.

Some people start using drugs or alcohol due to peer pressure or other social influences. They can be perfectly happy and metabolically healthy when they begin. Using excessive amounts of drugs and alcohol over time, however, impairs metabolism and mitochondrial function. Once impaired, people can get to a point in which they "need" the substances to feel normal. Notice that I used the word "normal" and not "good." Initially, when people begin using drugs and alcohol, they often feel good. This reinforces the behavior. People like feeling good. But over time, the brain adapts to these substances and tries to counteract them. As the brain changes, people can begin to feel "bad" when they aren't using the substance. This leads to a vicious cycle in which they now need the substance just to feel normal. They often can't get the same high that they once got. When they try to go without the substance, they suffer in some way. This usually drives them to use again. And now they are trapped.

Other people get into drugs and alcohol because they are already metabolically compromised. They already suffer from depression, anxiety, insecurity, psychosis, or something distressing. They are looking to feel better. If their symptoms are bad enough, they will try anything. As a broad overview, if they suffer from symptoms of underactive cells, such as some of the symptoms of depression, taking something that is stimulating will make them feel better. If they suffer from symptoms of overactive or hyperexcitable brain cells, such as anxiety or psychosis, taking something that is sedating and inhibits cell activity can make them feel better. If the substance works well, they can get hooked. In some ways, who can blame them? They just want to feel better. Sometimes, people don't necessarily feel "better" with substances;

they just feel "different"—numb or out of it. For some, this can be preferable to the way they feel otherwise. In any case, this is probably why *all* mental disorders are associated with higher rates of substance use disorders.

Drugs and alcohol can have immediate effects on mitochondrial function that can result in symptoms of many psychiatric disorders. Different drugs can quickly produce hallucinations, delusions, manic symptoms, depressive symptoms, cognitive impairment, and other symptoms. I can't review all of this in one chapter, but I'll share some highlights of how alcohol and marijuana can impact mitochondria.

Alcohol

Alcohol has profound effects on metabolism and mitochondria. When consumed in excess, it's known to be toxic to the liver and brain. Mitochondria play a primary role in this toxicity. I'll walk you through some of the science.

When people drink alcohol, the liver does most of the processing. An enzyme called *alcohol dehydrogenase (ADH)* converts it into *acetaldehyde,* a toxic molecule to cells. Another enzyme, *cytochrome P450 2E1 (CYP2E1),* can also make this conversion. CYP2E1 happens to be located directly on mitochondria or the endoplasmic reticulum. Yet another enzyme, *aldehyde dehydrogenase (ALDH),* then converts acetaldehyde into a less toxic molecule, *acetate.* ALDH comes in two forms: one that ends up in the cytoplasm and another that ends up inside mitochondria. Acetate then gets used by mitochondria as a fuel source. As you can see, mitochondria are playing a role in all of this.

If people binge drink, these enzyme systems back up, and acetaldehyde levels rise. The first signs of problems are with mitochondria. They swell up, have trouble producing ATP, and create more ROS. Numerous studies have documented mitochondrial impairment, and even destruction, from large doses of alcohol.[1] This is likely the cause of death in alcohol poisoning.

Chronic alcohol consumption leads to chronic oxidative stress, a sign of mitochondrial impairment. This leads to inflammation, which only makes the problems worse. All of this is occurring throughout the body, but particularly in the liver and brain.

Even short periods of heavy alcohol use can have lasting effects. Researchers looked at adolescent rats exposed to two weeks of binge drinking and the effects on mitochondria in their brains over time.[2] They found that binge drinking immediately impaired mitochondrial function, not surprising after what I just shared with you. However, the effects in the hippocampus lasted into adulthood, with reduced levels of mitochondrial proteins, reduced ATP production, and problems managing calcium.

Dr. Nora Volkow, the director of the National Institute on Drug Abuse, has been studying the connections between addiction and metabolism for years now and is a pioneer in this field. She and others have discovered some surprising findings in chronic alcoholics. When people drink alcohol, their brains use less glucose as an energy source and instead use acetate from alcohol.[3] Over time, alcoholics develop a problem with brain glucose metabolism. *Their brain cells become energy deprived when they are sober.*[4] When they drink alcohol again, acetate fuels these struggling brain cells and provides relief. This brain energy deficit may be one of the reasons that alcoholics have trouble staying away from alcohol. Volkow and others set out to see if they could help these struggling brain cells with something other than alcohol. They turned to the ketogenic diet.

They recruited thirty-three people with alcohol use disorder and admitted them to a detox unit.[5] Half of them got a ketogenic diet and the other half got the standard American diet for twelve weeks. The researchers detoxed participants using standard protocols and conducted a variety of blood tests and brain scans looking at brain metabolism in targeted regions. They found that people who got the ketogenic diet needed less detox medication and had fewer withdrawal symptoms. They also reported fewer cravings for alcohol. Brain scans showed improved brain metabolism and reduced levels of brain inflammation compared to those on the standard American diet. This pilot study demonstrated that a dietary intervention, seemingly unrelated to alcoholism, could make a big difference in the brains, and symptoms, of real people. This is how science can change the mental health field.

I want to point out one caution. As part of this work, the researchers tested what impact the diet might have on blood alcohol levels if a person were to drink. They tested rats on the ketogenic diet and found that their blood alcohol levels increased fivefold compared to rats on a normal diet, even though they all got the same amount of alcohol. This means that if people with alcohol use disorder were to try the ketogenic diet on their own, it could be dangerous if they drink. They could get much more intoxicated than usual. I don't mean to imply that interventions like this can't be used, but issues like this need to be considered, and people need to develop a safe way to manage these risks.

Marijuana

Marijuana is increasingly popular. Many people tout it as a "cure-all" for whatever ails you. It is thought to be good for seizures, pain disorders, nausea, anxiety, PTSD, and OCD. And yet, it can also *cause* mental symptoms, including learning and memory impairment, a lack of motivation, and possibly psychotic disorders.[6]

The brain energy theory offers a straightforward way to understand all these observations. They all relate to metabolism and mitochondria. The symptoms that improve are due to hyperexcitability. Any substance that reduces mitochondrial function in the correct cells could reduce these symptoms. However, such a substance could also *cause* symptoms if it impairs mitochondrial function too much. Is there any evidence that marijuana impacts mitochondria in these ways? Well, by now, you know I wouldn't ask that question if the answer wasn't a resounding *yes*.

Marijuana affects the endocannabinoid system in the human body. Receptors for cannabinoids are found throughout the body, but they are highly concentrated in the brain. There are primarily two types of receptors—CB1 and CB2. These are located on cell membranes, but CB1 receptors are also located directly on mitochondria. Because there are different types of receptors located on a wide variety of cells throughout the body, it's not fair to say there is one universal effect on all cells. However, the predominant theme in neurons is that marijuana slows the function of mitochondria through CB1 receptors.[7] Brain imaging studies of almost eight hundred adolescents, some who used marijuana and others who didn't, showed that the regions of the brain highest in CB1 receptors showed "accelerated age-related cortical thinning" in the marijuana users, meaning that marijuana's effects on these mitochondrial receptors was likely the cause of these brain areas getting thinner.[8]

A mouse study published in *Nature* found that mitochondria in astrocytes have a direct role in mediating the effects of marijuana—they control the amount of glucose and lactate (energy sources) going to neurons.[9] This, in turn, has direct effects on social behaviors. All of this is mediated through the CB1 receptors on mitochondria. When these receptors were activated by THC (the active ingredient in marijuana), it led to a reduction in both mitochondrial function and energy sources going to neurons. It also led to social withdrawal behaviors. When the researchers removed mitochondrial CB1 receptors, THC no longer had the same effects. The mitochondria weren't impacted in the same ways; the energy sources going to neurons weren't reduced; and the social withdrawal behaviors didn't occur, even though the mice were still being exposed to marijuana, and CB2 receptors were available on cells.

Another study from *Nature* tried to determine what causes the memory impairment from marijuana use. The researchers were ultimately hoping to better understand how memory works. The CB1 receptors on mitochondria again play a key role. The researchers found that marijuana's effect on CB1 receptors directly impacted mitochondrial movement, synapse function, and memory formation. When they deleted the CB1 receptors, marijuana no longer had any of these effects, and memory wasn't impaired. These researchers concluded: "By directly linking mitochondrial activity to memory formation, these data reveal that bioenergetic processes are primary acute regulators of cognitive functions."[10] In other words, brain energy and mitochondria play a primary role in our ability to remember.

There are many other addictive substances that impact metabolism and mitochondria, but I hope these two examples give you an idea of how substance use fits into the theory of brain energy.

Drug and Alcohol Treatment

Drug and alcohol treatment programs play a powerful role in improving mental and metabolic health. Reducing or stopping the use of substances that impair mitochondrial function is critically important.

Entire books are written on this topic. I won't try to review their conclusions here. Numerous strategies are available, including inpatient detox, residential programs, outpatient therapy, group therapy, medication-assisted treatments, twelve-step programs, and halfway houses.

Interestingly, a new area of research is on the use of psychedelics as a *treatment* for some psychiatric disorders. I'll get to that in Chapter 18.

Summing Up

- Drugs and alcohol affect your metabolism and mitochondria.
- Withdrawing from them can also impact metabolism and mitochondria in different ways.
- It's important to assess your use of substances, including tobacco, alcohol, caffeine, supplements, marijuana, and recreational drugs. These might be playing a role in your metabolic and mental health.
- If you are using any of these heavily, it could be an important contributing cause to any metabolic or mental symptoms you are having. You will likely need to address this before trying other interventions. If you have trouble doing this on your own, consider seeking professional help.

Chapter 17

CONTRIBUTING CAUSE

Physical Activity

Exercise is good for health. Many studies show that people who exercise are less likely to develop metabolic disorders, such as obesity, diabetes, and cardiovascular disease. This is one reason that exercise is so strongly recommended.

The same is true in relation to mental health. A study of 1.2 million Americans found that those who exercised had 43 percent fewer days of poor mental health, even after controlling for physical and sociodemographic characteristics.[1] This study found that *any* type of exercise was better than no exercise, but that the largest benefits were seen for team sports, cycling, and aerobic and gym activities. The optimal "dose" was forty-five minutes, three to five times per week.

Most people stop there. That's enough information to make a recommendation. If someone exercises for forty-five minutes, three to five times per week, that should solve the problem.

I truly wish it was that simple, but it's not. I've seen many patients who exercise regularly and still have crippling schizophrenia or depression. I want to explore the nuances of being active. Giving simple explanations with simple answers isn't going to address our mental health problems. If people follow the "forty-five minutes three to five times a week" advice and don't see results, they give up in frustration and disappointment. Don't worry if you're an exercise advocate—I'm still going to recommend exercise in the end.

The first thing to highlight was that the study with 1.2 million people was a correlational study. As you know by now, correlation doesn't equal causation. It's possible that people who exercise already have good mental and metabolic health, and that allows them to exercise. This would be *reverse causation.*

To illustrate how complicated this can all be, I'll tell you about another study. This one followed 1,700 women at midlife for twenty years to see if exercise prevents cognitive decline.[2] Most people would assume it does. However, after controlling for socioeconomic characteristics, menopause symptoms, hormone therapy use, and presence of diabetes and hypertension, they found that exercise didn't make any difference in cognitive symptoms. They concluded that "physical activity observed in later life may be an artifact of reverse causation." So, the news headline for this study is that exercise doesn't prevent cognitive decline. But this isn't so clear-cut. They "controlled" for diabetes and hypertension, as though exercise is independent of these variables. We know that it's not! Exercise decreases the likelihood of both, and may thereby decrease the risk of cognitive decline, too. We know they are all interrelated. And yet, some researchers and academic journals assume they aren't.

Exercise has been studied as a treatment for mental disorders, with depression being the most studied illness. The results are mixed, with some studies showing benefits and others not. A 2017 meta-analysis looking at exercise as a treatment for major depression included thirty-five studies with almost 2,500 participants.[3] Their conclusion: "Trials with less risk of bias suggested *no antidepressant effects of exercise* and there were no significant effects of exercise on quality of life, depression severity or lack of remission during follow-up." That's disappointing!

And yet the World Health Organization begs to differ. They issued a report in 2019 entitled *Motion for Your Mind.* They summarized their findings as follows: "A review of the evidence for the benefits of physical activity for people with depression, schizophrenia and dementia indicated improved mood, slowed cognitive decline, and delayed disease onset . . . "[4]

So, what are we supposed to believe? Does exercise help or not? It's tempting to err on the "safe" side and tell everyone to exercise, but if it doesn't actually work, that just sets people up for failure and makes the person recommending it seem less credible.

In otherwise healthy people, it's known that exercise can improve metabolic health. It's known to induce both mitochondrial biogenesis and mitophagy—the two things we are looking to do. This occurs not only in muscle cells but also in brain cells. Increasing mitochondria in brain cells should be helpful. So why don't treatment studies consistently show a benefit?

One reason may be insulin resistance. A study published in *Cell* found that it may block the benefits of exercise. Researchers had thirty-six people with varying levels of insulin resistance exercise and measured a plethora of biological measures before and after. They found that there were significant differences in energy metabolism, oxidative stress, inflammation, tissue repair, and growth factor responses, with most of these beneficial processes being *dampened or even reversed* in those with insulin resistance.[5] As I've already discussed, many people with chronic mental disorders have insulin resistance, so this may explain at the cellular level why exercise might be harder for them and why it might not work.

I suspect that the more important issue is that many people are taking substances and/or have lifestyle factors that impair mitochondrial function, and these are interfering with the beneficial effects of exercise.

Athletes, trainers, and coaches have long known that it takes more than just exercise itself to improve performance. All the factors that I discuss in this book play a role. If someone wants to improve their physical performance through exercise, they also need to pay attention to eating a proper diet, getting good sleep, and refraining from alcohol and drug use, among other things. For example, as I already discussed, alcohol can damage mitochondria and prevent mitochondrial biogenesis and mitophagy. This is why we've all heard the advice that if you're training for an important athletic event, or even if you just want to lose weight, you need to stop drinking. Improving metabolism involves many lifestyle factors in combination, not just one.

Medications can also play an adverse role. In theory, any medication that impairs mitochondrial function might prevent exercise from working. One study looked directly at this issue for the commonly prescribed diabetes medication metformin. Researchers had fifty-three older adults participate in twelve weeks of aerobic training and assigned half of them to take metformin and the other half a placebo. Both groups experienced some benefits from the exercise, such as a reduction in fat mass, glucose, and insulin levels. However, improvements in muscle mitochondrial function were abrogated in the people taking metformin. The metformin group had no overall change in whole-body insulin sensitivity, even though the placebo group experienced improvement. These researchers summarized their findings in the title of their study: "Metformin Inhibits Mitochondrial Adaptations to Aerobic Exercise Training in Older Adults."[6] So, in all the studies looking at the effects of exercise for weight loss, diabetes, or mental disorders, we would need to know if any of those participants were taking metformin. If they were, they were likely being set up for failure to improve their mitochondrial function, and this might be the reason some of the studies showed no benefit.

Metformin is one of the "mildest" diabetes medications with the fewest side effects. Many other diabetes medications, including insulin itself, cause weight gain and even more insulin resistance over time. But it's not limited to diabetes medications. As you now know, some psychiatric medications, especially the antipsychotic medications, are known to cause serious metabolic disturbances and mitochondrial dysfunction. People taking any of these medications will likely fail to get the full benefits of exercise.

The research on exercise for mental illness didn't factor any of this into their studies.

Mitochondria play a direct role in translating exercise into beneficial effects in the brain. When people exercise, one of the benefits is that they usually develop new neurons in the hippocampus from stem cells. This process has been found to be directly related to both mood and cognitive disorders. The growth of these stem cells into new neurons depends upon mitochondria. When researchers genetically manipulated mitochondria to inhibit or enhance their function, the development of these new neurons was inhibited or enhanced, respectively.[7] Based on this research, it appears likely that if someone has poor mitochondrial function in this brain region, they may not get the same benefits from exercise that other people get. However, if we can restore their mitochondrial health, it's possible that we can change this.

As a rule of thumb, exercise can do one of two things: It can help people maintain their current abilities, or it can improve their abilities. This translates into maintaining your current metabolic status or improving it.

Taking a leisurely walk around the block helps people maintain their current metabolic status. This is useful. It's certainly better than losing strength or ability. However, to improve metabolic capacity, people need to push themselves. They must work toward getting faster, stronger, more flexible, doing more reps, or achieving some other metric of increased capacity. We know that when this happens, the number of mitochondria in their muscle and brain cells increase, and the health of those mitochondria improve as well.

One of the challenges with exercise is that asking people who are metabolically compromised to push themselves involves risk. There are risks of injury and even heart attacks. Therefore, exercise needs to be managed in a safe way. Physical therapists, personal trainers, and others will play an essential role for some people.

The bigger challenge is getting people who are metabolically compromised to follow through with an exercise routine. They lack energy and motivation. Their metabolism is doing this to them. It's not their fault.

Overcoming this inertia can be difficult. Nonetheless, it *can* be done with support, encouragement, and education.

Exercise as a Treatment

Should everyone exercise? I would say yes. But it's important to keep in mind that it will be much harder for people with chronic mental disorders, and they may not notice benefits right away. It's also important to take an inventory of all the factors that might be impairing your mitochondria and metabolism; reducing or eliminating these will allow exercise to work.

Nonetheless, exercise won't be the answer for everyone. As I've been discussing, there are many factors that play a role in metabolic and mental health. Exercise is only one of them. For people with vitamin or hormone deficiencies, for example, exercise isn't going to solve the problem, but it certainly won't hurt.

Summing Up

- Exercise can play a role in *preventing* mental and metabolic disorders.
- Exercise may be more difficult if you have insulin resistance or any condition associated with mitochondrial dysfunction. It may take longer to show benefits. This doesn't mean it won't work; it just means that you should try to be patient and not expect immediate results.
- Identifying, removing, and/or reducing substances and lifestyle factors that impair mitochondrial function will be required to realize the full benefits of exercise. These can sometimes negate the benefits of exercise.

- Exercise can be an effective treatment for some people with mental disorders, but for others, it may not be the solution.
- People with injuries or physical limitations should work with their healthcare provider to implement an exercise program safely. This may include working with a physical therapist.
- Even if exercise doesn't improve your mental symptoms, you should still exercise, as it offers numerous other health benefits. Humans are meant to move.

Chapter 18

CONTRIBUTING CAUSE

Love, Adversity, and Purpose in Life

Metabolic and mental health require a combination of both biological and environmental factors. I've told you about many of the biological factors. The environment includes many things—food, shelter, temperature, light, infections, allergens, and lifestyle choices—some of which we've covered. But it also includes people, experiences, love, and purpose in life. Although most people see these latter constructs as psychological and social issues and often assume they aren't related to biology, they actually play profound roles in metabolism. They are all interrelated and inseparable. We adapt and respond to our environments, for better or worse.

Use It or Lose It

The phrase "use it or lose it" is usually associated with exercise and muscle strength. When people use, or stress, certain muscles, they become larger and more resilient. Not only do they grow in size, but they also develop more mitochondria. This is true even when the size of the muscles isn't that much larger. For example, some long-distance runners can be very thin. Their muscles aren't all that big, yet they contain more mitochondria than the muscles of people who don't run. These mitochondria give their muscles the endurance they need to go long distances.

Not using muscles results in atrophy, or shrinking. This can be striking when people break a bone and are in a cast for several weeks. Their muscles shrink rapidly. Why? When the body isn't using something, it diverts metabolic resources from it. The body is always adapting and adjusting. It spends its energy wisely. If muscles aren't being used, they don't get much glucose or as many amino acids. They quickly shrink. The good news is that the body keeps a memory of what those muscles once were. Once the cast comes off, the muscles quickly return to their normal size if they get used in the same ways again. This truly depends on what size they were beforehand. Big bodybuilders will quickly regain their massive muscles, while frail, elderly people will only get a tiny amount of muscle growth.

This concept of "use it or lose it" applies to more than just muscles. It also applies to the brain. The best evidence for this comes from studies of children as their brains are developing.

Some human skills and traits need to be acquired at the right times. The brain undergoes "developmental windows" during which it is ready to learn and adapt. However, the "environment" must offer experiences needed to acquire these skills or else they can be altered for life. Social skills are one example.

Social skills are important to human survival. They allow us to live in families, towns, and societies. They require two things to develop properly: (1) normal brain development to acquire and store the information and (2) learning experiences from other humans. If either is absent, problems will be obvious. The biology side can be understood in terms of brain development, mitochondria, and metabolism, as I have already discussed. The environment side depends upon parents and caretakers primarily. There is a tremendous amount of literature looking at the effects of attachment, neglect, abuse, and social deprivation on human development. Many of us would say these things relate to *love*, or a lack thereof. They play a profound role in human development, including the acquisition of social skills. Children deprived of appropriate learning opportunities often lack the skills needed to navigate the world effectively. In extreme cases, the consequences can be catastrophic.

The findings of researchers studying Romanian orphans exemplifies how tragic this can be. The orphanages where these children were housed were profoundly neglectful, and the children who experienced this neglect were found to suffer from a range of diagnostic categories including autism, learning disorders, mental retardation, PTSD, anxiety disorders, impulse control disorders, mood disorders, personality disorders, and even psychotic disorders. Once again, numerous diagnostic categories, not just one. Their brains were deprived of the appropriate opportunities to learn how to be "human" in society and the consequences were sometimes devastating. The malnutrition, stress, and trauma they experienced undoubtedly also played roles, but so did the lack of appropriate learning experiences.

These children's brains didn't develop normally. If areas of the brain that perform certain functions aren't being used, they don't grow and thrive. One research group studied metabolic brain scans of ten such children and compared them to normal control and even epileptic children.[1] Sure enough, they found widespread areas of reduced brain glucose metabolism, indicating a brain energy problem in these former Romanian orphans. Sometimes, these deficits can be corrected later in life; but in some cases, they appear to be permanent. Developmental windows can close, and the opportunity for normal brain development can be lost forever.

It's not always so extreme. For example, children exposed to more screen time are more likely to develop ADHD. There are two ways to interpret this observation. One explanation is that the environment is driving the subsequent diagnosis of ADHD. These children are learning that constant stimulation is the norm given the content on the screens. They are not learning patience, focus, and concentration even if their brains are ready to learn. These developing brain networks will receive fewer metabolic resources since they aren't being used, much like the unused muscle. They may not develop normally, or they may not be as strong and robust as they could otherwise be. This can result in ADHD symptoms. However, it's also possible that this could be due to reverse causation, and biology might be the problem. If these children have inadequate metabolism in specific brain regions, they may be unable to focus. This may drive them to use screens as a source of entertainment. If this explanation is correct, then correcting the metabolic issue will be the first step in solving the problem.

The concept of strengthening brain regions is obvious to most of us through clichés like "practice makes perfect." This applies to learning a new language, playing basketball, or learning to play the piano. When we use our brains in specific ways, neurons grow, adapt, and form new connections. If we use them, they grow. If we don't, they wither. This all relates to metabolism and mitochondria. They adapt to our needs.

Stress

I now return to stress. I have discussed this throughout the book and have already told you how it plays a powerful role in mental and metabolic health. I'll review some highlights, share some new information, and then get to treatments.

Recall that the stress response requires energy and metabolic resources. These resources are being diverted from other cells throughout the brain and body, and these other cells can suffer. For example, if a young boy is chronically stressed, he will have more difficulty learning in school. It's not necessarily because he isn't trying. The stress response is taking energy that could otherwise be used for brain functions like focus, learning, and memory.

Stress puts cellular maintenance functions on hold. If it occurs for a prolonged period, it can result in maintenance problems in cells, especially ones not being used much, which can lead to symptoms of mental and metabolic disorders. Any cells that are already metabolically compromised can begin to malfunction under stress, which can exacerbate symptoms of mental and metabolic disorders.

In Part Two, I talked about how mitochondria play a critical role in the stress response. They influence all aspects of the stress response, including the production and regulation of key hormones and neurotransmitters, nervous system responses, inflammation, and epigenetic changes. When mitochondria are not functioning properly, all of these can be affected.

One research study demonstrated a direct relationship between everyday stress and changes in mitochondrial function in humans.[2] The researchers developed a test of mitochondrial health that includes both the quantity and function of mitochondria within white blood cells and assessed whether this metric was associated with daily stress. They studied ninety-one mothers, some who had children with autism and others who had neurotypical children, and assessed their daily mood and stress levels to see if these were related to the mitochondrial health index (MHI). They found that they were. Overall, the mothers who had high levels of stress and low moods had lower MHI. But of course, stress levels and moods can change on a daily basis. The researchers looked at this specifically. When the mothers were in a positive mood, the MHI went up afterward, sometimes within one day. In other words, the health and function of mitochondria in white blood cells were changing in response to the mothers' daily moods and stress levels. This research demonstrates how stress can lead to impairment in mitochondrial function, which can then affect overall health.

All humans experience stressful life events. In the 1960s, Dr. Thomas Holmes and Dr. Richard Rahe, both psychiatrists, studied five thousand medical patients to see how stressful life events can contribute to physical illness. They identified some common life events and ranked them by how much they impacted overall health. The Holmes-Rahe Stress Inventory is still available today and can give you a sense of what life events are most stressful. Some top ones include the death of a spouse or close family member, divorce, personal injury, getting fired, and even retiring. These select ones involve some type of loss—losing someone important to you, losing your health, or losing your job (even voluntarily). What makes these so stressful? There are many reasons, and they can be different for different stressors, but one common theme is that they all relate to purpose in life.

Purpose in Life

Humans are driven to have a sense of purpose. I believe this is hardwired into our brains, given that this single construct has been highly associated with both metabolic and mental health. When people lack a sense of purpose, it appears to induce a chronic stress response and can lead to many poor health outcomes. Purpose in life is multifaceted, however. It usually includes many things, not just one. The stress inventory that I just mentioned highlights three possibilities: relationships, taking care of yourself and staying healthy, and having a job.

Dr. Viktor Frankl, an Austrian psychiatrist who was taken prisoner by the Nazis during World War II, deserves credit for highlighting the powerful role of meaning and purpose in life. In his book *Man's Search for Meaning,* he described his observations of the other prisoners in the concentration camp. Most became severely depressed, for obvious reasons. However, some did not. Some of the prisoners appeared to hold on to hope that they might live and escape. Frankl argued that the common denominator among them was that they all had a sense of purpose in life: they had a reason to fight and try to stay alive.[3] Frankl went on to develop a psychotherapy, *logotherapy,* based on the construct of meaning and purpose in life. Many of its tenets are still embedded in current mainstream psychotherapies.

The concept of purpose in life continues to be studied today and has been highly correlated with a wide range of metabolic and mental health outcomes. It's not surprising that a low sense of purpose in life is associated with depression, given that depression itself might make people feel this way. It may just be circular logic. However, a lack of purpose is also associated with metabolic disorders and even longevity, consistent with the brain energy theory. For example, one study of nearly seven thousand US adults ages fifty-one to sixty-one found that those who had the lowest sense of purpose in life were about 2.5 times more likely to die an early death than those with a strong sense of purpose in life.[4] They were dying of heart attacks, strokes, respiratory disorders, and gastrointestinal conditions. These researchers noted other studies showing that a strong sense of purpose leads to lower levels of cortisol and inflammation, which might explain these health benefits. A 2016 meta-analysis of ten prospective studies that included over 136,000 participants also found that having a higher sense of purpose in life was associated with reduced all-cause mortality and cardiovascular events.[5]

When discussing purpose in life, it's important to include spirituality and religious beliefs. For many, these play a powerful role in how they understand their existence. Researchers have studied the effects of religious beliefs and practices on a variety of health outcomes and, in general, have found many beneficial effects. For example, one study looking at adults at high risk for depression found that those who reported religion or spirituality to be highly important were 90 percent less likely to develop depression compared to those who reported them of low importance.[6] The researchers conducted brain scans on these people and found differences in the thickness of certain brain regions based on how important religion and spirituality were to the participants. These brain differences may explain the protection against depression. The Nurses' Health Study followed almost ninety thousand women for more than fourteen years and found that women who attended religious services at least once per week were five times less likely to commit suicide than those who never attended religious services.[7] A systematic review of religious beliefs and practices and their effects on health found a beneficial relationship with several other health outcomes, such as decreased rates of cardiovascular disease and all-cause mortality.[8] One challenge with this type of data, however, is the possibility of reverse causation again—it might be that people who are already healthy are able to attend religious services and connect with people there, but people who are already depressed or metabolically compromised aren't. Despite this possibility, I'll soon share with you data that suggests some religious practices, such as meditation and ritual prayers, might directly help improve metabolism and mitochondrial health, so these may, in fact, play a causal role in improving health outcomes.

For those who don't believe in God, I don't mean to imply that you need to start believing in God to improve your health. I share this information because it relates to purpose in life and has been found to influence both metabolic and mental health. There are other ways to find purpose that can play just as powerful a role.

Addressing Love, Adversity, and Purpose with Treatment

All of this can play a role in treatment. First, it's important to point out a general rule of thumb for human health—people need to develop and maintain full lives, ones that include what I call the four Rs: close *relationships*, meaningful *roles* in which they contribute to society in some way, adherence to *responsibilities* and obligations (not just to the people in one's life, but to society as a whole, such as not breaking laws), and having adequate *resources* (money, food, shelter, etc.).

Many societal factors can interfere with people's abilities to achieve all of this, including war, trauma, poverty, malnutrition, neglect, racism, homophobia, misogyny, all the ACEs, and so many others. Efforts are needed to address these social problems, for as long as these continue, mental illness will continue to exist. However, people who have been affected by these barriers and atrocities can still recover. They can use science-based approaches to understanding and addressing the impact of these experiences on their metabolism and mitochondria. I'm hopeful this book and the theory of brain energy will help at least some of them.

Psychotherapy

Addressing the myriad of psychological and social factors that can impact metabolism is an important part of treatment. Friends, family, coworkers, teachers, mentors, or people in the community can often help. Some people, however, will need professional help in the form of psychotherapy.

There are countless books and academic articles on how psychotherapy can improve mental health. I won't even try to review all the research. Instead, I'll share just a few of the benefits and some possible reasons why it works:

- Psychotherapy can help people *resolve conflicts* with other people and with their roles in life. When people are unable to do this on their own, it increases stress, which takes a toll on metabolism.
- Psychotherapy can offer specific *skills and strategies* to reduce stress and cope with symptoms, which can improve metabolism broadly.
- Psychotherapy can help people *change behaviors*. Cognitive behavioral therapists have long known that changing behaviors can sometimes result in changes in thoughts and feelings. Clinicians treating eating disorders or substance use disorders are often focused on changing behaviors. Addressing behaviors to enhance sleep can produce benefits. As I've discussed, all these behaviors play a direct role in metabolism and mitochondrial function.
- Psychotherapy can help people *understand who they are and what they want from life*. This can help some people develop a sense of meaning and purpose, which impacts both mental and metabolic disorders.
- Psychotherapy can provide *new learning* to overcome maladaptive beliefs, behaviors, and responses. For example, when people are traumatized, they can sometimes overgeneralize the danger of things that they remember from that experience. Someone who associates certain music, clothes, or cologne with their abuser might be triggered by these everyday experiences. If the abuser is no longer a threat, this is not adaptive or helpful, even though it is understandable. One type of therapy, prolonged exposure, can

reduce the stress response to these triggers, which can improve metabolic health.

- Psychotherapy can *"exercise" underused brain circuits*. Remember "use it or lose it"? If a brain region is underdeveloped, some types of psychotherapy can help. Treatments that focus on empathy, relationships, social skills, or improving cognitive abilities can all strengthen brain circuits that have been underdeveloped. This assumes that these brain regions are metabolically healthy enough to learn and store new information. In some cases, they may not be. In these situations, a different metabolic intervention might be needed first. Once brain health is restored, however, "exercising" and restoring health to these brain regions will still be needed.
- Psychotherapy can simply offer a *relationship with a compassionate and helpful human being*. It has long been known that the "therapeutic alliance," or a good relationship between the therapist and client, plays a role in psychotherapy outcomes. We come back to one of the realities of human existence: We all need other people. We need relationships in which we can express ourselves and be ourselves. Without them, metabolic health can become compromised due to a chronic stress response. For those who don't have meaningful relationships, psychotherapy can provide one. Obviously, a goal should be to help these people develop sustaining relationships outside of therapy. However, this can take time. For some people, symptoms of their brain dysfunction can make it difficult to achieve.

Psychedelic Therapy

One emerging treatment related to all of this is the use of psychedelic drugs in psychiatry. Hallucinogens, such as psilocybin or "magic mushrooms," are receiving increasing attention as a possible treatment for depression, PTSD, and other conditions, with small pilot trials showing benefits. One research group explored how and why these work. They noted, "psychedelics can reliably and robustly induce intense, profound, and personally meaningful experiences that have been referred to as 'mystical-type', 'spiritual', 'religious', 'existential', 'transformative', 'pivotal' or 'peak.'"[9] They surveyed 866 users of psychedelics over time and found alterations in metaphysical beliefs that often persisted for longer than six months. These persistent metaphysical beliefs were correlated with improved mental health outcomes. This line of research suggests that psychedelics might work by connecting people with spirituality or God, or offering them a sense of meaning and purpose. I should point out that using these on your own is not recommended. The research trials are pairing the use of psychedelics with guided therapy sessions in order to maximize benefits. Using them on your own may result in a "bad trip" or could possibly even trigger a manic or psychotic episode.

Stress Reduction

Reducing stress is an important part of treatment. In addition to all the ways that psychotherapy and talking with other people can help reduce stress, there are two ways that people can do this on their own: (1) reduce or eliminate stressful environmental factors or (2) attempt to reduce your stress response when it is safe to do so.

The easiest way to manage stress is to reduce or eliminate stressors when possible. For some, this is a realistic goal. A highly demanding work or school environment that is overwhelmingly stressful can be managed. The employee can find a new job, or the student and parents can reduce course load, seek academic accommodations for a disability, or change schools to find a better fit. Creating a life that is manageable, pleasurable, and rewarding is something we should all strive to do.

When stressful life events occur, people *will* experience a stress response. This is normal and expected. But when the threatening situation is no longer a danger, reducing the stress response can have powerful, beneficial effects.

Stress-reduction practices have been used for millennia. Some of these are not usually thought of as "stress-reduction techniques," but instead as longstanding religious practices, such as meditation, praying, and chanting. Additional practices include yoga, Pilates, tai chi, qigong, mindfulness, and breathing techniques. Many of these interventions have been shown to improve both mental and metabolic health. I won't review all the interventions and health conditions, as there are many. However, I will share a couple of studies directly linking these benefits to metabolism and mitochondria.

Researchers from Harvard Medical School had long known that the relaxation response can play a powerful role in both mental and metabolic health. The relaxation response, or RR, is a term they use to describe any of the stress-reduction techniques I already mentioned, such as meditation. Studies have demonstrated improvement in hypertension, anxiety, insomnia, diabetes, rheumatoid arthritis, and aging itself. They set out to better understand how this intervention works. They recruited nineteen healthy, long-term practitioners of daily RR practice, nineteen healthy controls, and twenty people who recently completed eight weeks of RR training. They took blood samples from all of them and looked at differences in gene expression. For those doing RR, they found significant differences in genes related to "cellular metabolism, oxidative phosphorylation, generation of reactive oxygen species and response to oxidative stress." As you now know, these are directly related to mitochondria.[10]

In a follow-up study, the researchers recruited twenty-six people who had been practicing regular RR techniques for four to twenty years and another group of twenty-six people who had never regularly practiced RR but were willing to complete eight weeks of training.[11] All the participants were then asked to listen to a twenty-minute RR recording and, in a different session, a twenty-minute health education recording. Blood samples were taken before, immediately after, and fifteen minutes after listening to each recording, and these samples were analyzed for gene expression. The researchers found that "RR practice enhanced expression of genes associated with energy metabolism, mitochondrial function, insulin secretion and telomere maintenance, and reduced expression of genes linked to inflammatory response and stress-related pathways." A specific mitochondrial protein (mitochondrial ATP synthase) and insulin were the top two upregulated molecules. These researchers concluded, "Our results for the first time indicate that RR elicitation, particularly after long-term practice, may evoke its downstream health benefits by improving mitochondrial energy production and utilization and thus promoting mitochondrial resiliency . . . " As you now know, this is precisely what we are trying to do to improve mental and metabolic health!

Rehabilitation Programs

Many people with chronic mental disorders lack the skills needed to independently survive and thrive in society. Some don't know how to make friends. Others don't know how to manage a daily schedule. Still others don't know how to hold a job. Many feel they lack purpose in life.

Their symptoms prevent most of these people from doing these things. Even if they learned how to do them prior to their illness, they might now be out of practice. For those who developed their illnesses at a young age, they may never have learned these skills in the first place.

Restoring metabolic health to their brains won't automatically teach them everything they need to know. They need training and practice. It's like rehabilitation after a sports injury. You have to first restore the function of the muscles, bones, ligaments, or tendons, but then the person must also practice and build up strength again. Without this practice, abilities won't be restored.

There are currently rehabilitation programs that offer education, job training, and basic life skills to people with chronic mental disorders. Unfortunately, current research suggests they aren't very effective. This is likely because brain function was not restored first. If people are trying to do tasks when their brains aren't working properly, they are being set up for failure. It's like an athlete trying to run a marathon with a torn ligament. However, if we can restore proper brain function, then rehabilitation has a decent chance of working. The goal is to help people live as productive members of society. Many of them have been beaten down for years and may feel hopeless about this goal. This hopelessness also needs to be addressed.

In all of this, compassion from other humans is required. Job programs and reentry into society are needed. These people have to find reasons to live. They need to feel useful. They need to feel respected. All of this requires the involvement of other people.

Summing Up

- Our environment and experiences play a critical role in our metabolic and mental health.
- Close relationships are important to human health.
- Everyone should strive to have at least one role in society that allows them to contribute and feel valued. This can take the form of being a student, employee, caretaker, volunteer, mentor, or other role. It can be as simple as having household chores.
- Psychotherapy can play an important role in your metabolic treatment.
- Psychotherapists can add many new tools to their arsenals using the brain energy theory. They can assist people in implementing metabolic treatment plans, which may include diet, exercise, exposure to light, and all the other possibilities mentioned in this book.
- People with chronic mental disorders may have a lot of catching up to do if they restore their brain health. Their full recovery will require rehabilitation, job training, and other programs.
- Society needs to work together to ensure that everyone has adequate relationships, roles, resources, and responsibilities. All humans are not created with equal abilities, but that doesn't mean that all humans can't contribute, be safe and secure, and live meaningful lives. Compassion and kindness are essential to this process.

Success Story: Sarah—Exercise and Finding Her Purpose

Sarah was seventeen years old when I first met her. She had been diagnosed with ADHD and a learning disability in the eighth grade, had suffered from anxiety and insomnia for as long as she could remember, and began having panic attacks at age fourteen. She was also depressed and had low self-esteem. She struggled in school, despite taking medications for her ADHD, and she had few friends. She had a strong family history of mental illness, with her mother, brother, sister, grandmother, two uncles, and an aunt all having been diagnosed with depression, anxiety disorders, and/or bipolar disorder. That didn't bode well for her long-term outcome. She had already tried eight different medications, which helped with her ability to concentrate, but also came with side effects and continuing symptoms. She would sometimes get so depressed that she would stay in bed all day. On top of her mental symptoms, she also had migraine headaches and frequent stomach pains.

She got into college and did her best, but she struggled. Her family expected her to finish college, and this stressed her even more. She often felt that she could never live up to people's expectations. She tried more antidepressants, but they didn't help.

Things changed when she decided to take Pilates classes. She loved them! She began exercising regularly and noticed that many of her mood and anxiety symptoms were getting better. At age twenty-three, she got a job as an instructor at the studio—she was now exercising almost every day for hours per day. That was the game-changer for her. She came to an appointment after about two months on the job saying, "I feel great! I've never felt this good in my entire life." On top of the exercise, she had become passionate about helping others improve their health, had made new friends, and now had a boyfriend. Despite her parents' expectations that she finish college, she decided to drop out and focus on exercise as her career. That was almost ten years ago and Sarah continues to do well to this day. She still takes stimulant medication for her ADHD symptoms, but was able to stop all the other medications. After tapering off them, she said she actually felt even better.

Sarah's story highlights the power of exercise as a metabolic treatment, but also the psychological and social aspects of finding meaning and purpose in life, stress-reduction practices, having a social support network, and not letting the expectations of others rule us. We are all different, and want—and need—different things. Sarah found her own way to metabolic and mental health.

Chapter 19

Why Do Current Treatments Work?

What do talking, chemicals, electricity, magnetic fields, and brain surgery have in common? They are all evidence-based treatments for mental disorders! So, why do they work? They all impact metabolism and mitochondria.

I've already addressed medications and psychotherapy in earlier chapters. Now, I'd like to briefly explain how I think about these other treatments in the context of the brain energy theory. If this theory is true, there should be plausible explanations for why these treatments work, at least for some people.

ECT and TMS

Electroconvulsive therapy (ECT) and transcranial magnetic stimulation (TMS) are effective interventions for a wide variety of mental disorders. For some conditions, such as severe depression or catatonia, ECT is considered the gold standard and most effective treatment that we have available. Why do they work? The field doesn't currently offer a comprehensive explanation. Changes in neurotransmitter levels and hormones are thought to play a role, as well as increased neuroplasticity. The brain energy theory offers one comprehensive explanation.

Electricity from ECT and electromagnetic energy from TMS deliver energy directly to the brain. There's probably no better example of a treatment being related to "brain energy." This energy stimulates mitochondria, and in turn, mitochondrial biogenesis. When we push ourselves in exercise, the body senses that it needs more capacity, so it produces more mitochondria to deliver this capacity. ECT and TMS appear to do the same thing. This can improve neurotransmitter and hormonal imbalances, as well as increasing neuroplasticity. These findings can be understood through mitochondria.

The direct effects of ECT on mitochondria have not been extensively studied. However, one group of researchers did demonstrate increases in mitochondrial activity of the hippocampus, striatum, and cortex of rats after delivering ECT.[1] Another group found increased mitochondrial biogenesis and enhanced synapse formation in the hippocampus after just one ECT treatment.[2] They also found that a series of ten treatments resulted in lasting improvements in the number of mitochondria and synapse formations three months later.

TMS has been shown to improve oxidative stress, reduce inflammation, increase neuroplasticity, and impact neurotransmitter levels.[3] As you know, these all relate to mitochondrial function. The evidence for the direct effects of TMS on mitochondria, however, is also sparse. One study found an increase in ATP levels in a rat model of stroke.[4] Another study found increased mitochondrial integrity after TMS treatment, also in a rat model of stroke.[5]

Interestingly, psychiatry isn't the only field that uses electricity to fix metabolic problems. Cardiology commonly uses cardioversion (or shocking of the heart) when the heart is failing metabolically. Sometimes it just needs a jump-start.

Brain Surgery and Electrical Stimulators

Brain surgery is sometimes used as a last resort in people with chronic, debilitating mental disorders. Sometimes it can help. Why?

This is fairly straightforward. If an area of the brain is overactive due to hyperexcitable brain cells, cutting it off from the rest of the brain can decrease symptoms. This is commonly done in epilepsy treatment. The same applies to hyperexcitable brain regions causing mental symptoms.

In other cases, some brain surgeries implant electrodes to stimulate cells. This, too, is straightforward. It's a way to stimulate underactive brain regions. It's used in cardiology when pacemakers are implanted to deal with reduced function of the pacemaker cells of the heart. It works much the same way for underactive brain regions. Paradoxically, fast-paced stimulators can sometimes *suppress* overactive brain regions.

Electrical stimulators have also been applied to the vagus nerve. This is called vagal nerve stimulation, or VNS. This has been helpful for epilepsy and depression, and is also being studied for PTSD, Alzheimer's disease, schizophrenia, OCD, panic disorder, bipolar disorder, and fibromyalgia.[6] Again, one treatment for so many seemingly unrelated conditions. *The brain energy theory* connects them all.

Summing Up

For severe, treatment-resistant conditions or in life-threatening emergencies, ECT, TMS, VNS, and/or brain surgery can all play a role in treatment. However, the brain energy theory offers numerous other treatment options before these might be needed.

Chapter 20

Putting It All Together

DEVELOPING YOUR METABOLIC TREATMENT PLAN

When a flower doesn't bloom, you fix the environment in which it grows, not the flower.
—Alexander den Heijer

The above quote is a powerful metaphor for addressing problems with metabolism and mitochondria. In most cases, there is not a "defect" in the person but in the environment. "Fixing" a mental illness requires identifying the problems and addressing them. The "environment" in this case is writ large. It includes all factors that affect metabolism and mitochondria, such as diet, exercise, stress, light, sleep, hormones, inflammation, relationships, love, and meaning and purpose in life, to name just a few. Yes, some people may have inherited epigenetic factors, such as micro-RNAs, and these might be a contributing cause to their mental illness, but these can be changed, too. Metabolism is malleable, and there are many ways to improve it.

You will recall that I compared all mental illnesses to delirium. Each and every case of delirium is different, even though the symptoms can be similar and the diagnosis the same. Solving the problem of delirium requires medical detective work to understand what is causing it. Oftentimes, it's more than one thing—a perfect storm in which many assaults on mitochondria are occurring at the same time. They all need to be identified and dealt with. This applies to every case of mental illness as well.

The brain energy theory supports existing treatments for mental disorders. These will continue to play an important role. However, it also calls for radical changes. To address problems with metabolism, comprehensive approaches are usually required. Sometimes, identifying a simple problem and delivering a single treatment can work. Vitamin and hormone deficiencies are examples, and sometimes, simply taking an antidepressant can do the trick. But unfortunately, more often than not, simple solutions are usually not the answer.

This goes against a pervasive message that we hear on a daily basis. Everyone wants simple solutions. We see on television every day that we can fix our problems with a pill. We just need to talk to our doctors and get those new prescriptions. If one pill doesn't do it, by all means, take more. We hear the same messages from diet gurus and health experts. Don't eat fat, and you'll lose weight. Take this vitamin or supplement, and that will fix things.

These messages, of course, are alluring. All we need to do is take a pill or do a simple intervention to fix our problems. They are much more appealing than doing the complicated work of figuring out what's actually wrong, and then correcting the problem or problems, which might include making lifestyle changes. In reality, the simplistic fixes usually don't work, at least not fully or permanently. The skyrocketing rates of mental and metabolic disorders are a clear testament to the failure of this approach. The field of medicine is increasingly recognizing this with its push toward *personalized medicine,* which acknowledges that there are many pathways to illness and that one-size-fits-all solutions often don't work. People need unique treatment plans tailored to their individual situations and requirements.

Working with a Clinician

When treating serious mental disorders, it's critically important that you work with a competent clinician. Serious mental disorders can be dangerous, and people should not expect to treat these without help. "Serious" symptoms include things like hallucinations, delusions, suicidal thinking or behaviors, self-injury, aggression, out-of-control substance use, severe eating disorders, and other dangerous behaviors. These are not do-it-yourself projects to handle on your own at home. You deserve competent and compassionate medical care, so please get help developing and implementing your metabolic treatment plan. It's important that you have support and expertise to get to a healthy and safe place as quickly as possible.

For those with chronic disorders, even if just mild or moderate, you may also need to work with a clinician. A thorough medical evaluation might reveal factors that are playing a role in your illness.

Where to Start

All the contributing causes and interventions I have discussed are interrelated and affect each other. If one factor is off, others will be, too. For example, if your sleep is off, your eating behavior and use of substances might be as well. Even something like the gut microbiome can be affected by sleep, light exposure, and stress. So if your microbiome is off, changing these other contributing causes might correct that issue. Alternatively, changing your microbiome might improve your sleep or stress levels. Think of everything as being part of one or more feedback loops. Therefore, *all* the treatments outlined in this book are possible treatments for you, even if you don't recognize that you have a problem in these areas. Addressing mental symptoms might require changes in your sleep, diet, or light exposure.

In some cases, it won't be clear what's causing your metabolic dysfunction. Worry not. Treatment approaches can still work. The goal of treatment will be to use interventions that are known to improve mitochondrial function and/or increase the supply of mitochondria. In most cases, if our cells have enough normal and healthy mitochondria, they function properly. Mitochondria know what to do. They can often fix the problem on their own.

Some of you might be overwhelmed by all the treatment options. When trying to improve your metabolic health, recognize that success requires a *multifaceted approach that will take time*. But this means you don't have to do everything at once—nor should you. Start with one treatment, try it for a few weeks or months, and then add additional treatments as needed. Oftentimes, as your metabolism begins to improve, it gives you more energy and motivation. This makes it easier to make other changes. As people begin to feel better, they often surprise themselves with how much more they can accomplish. When people complete their "metabolic treatment plans," they often don't recognize themselves. Not only can they reduce their symptoms of mental illness, lose weight, or have more exercise endurance, but they usually also notice stress reduction, more self-confidence, more connections with people, newfound abilities, and other life-enhancing positives.

In most cases, *you* can decide what intervention you want to start with. Once you choose one, make it a SMART goal: Specific, Measurable, Achievable, Realistic, and Timely. Once you master one intervention, add another. Keep going until you achieve your desired results.

In some cases, however, one intervention may need to take priority over all others, as it could be having catastrophic effects on your metabolism. Two obvious examples are heavy substance use or living in an abusive environment. For those who are heavily using alcohol or drugs, the other interventions will likely be futile until the substance use stops, so address that first. For people in physically abusive relationships, the first step should be to focus on a plan to get out of that environment, as difficult and dangerous as that might be. People in this situation will likely need tremendous support from family, friends, or a domestic violence program. Trying other metabolic interventions without this first step will likely be futile to address their mental and metabolic problems.

All the interventions I've discussed have the potential to change your metabolism. As a rule of thumb, metabolic interventions can have four types of effects on the body and the brain:

1. **Initiation**—When you first make a change, this can abruptly shift your metabolism. Sometimes, this can be helpful. Other times, it can make things worse, at least initially.
2. **Adaptation**—Whenever you make a metabolic change, your body will work to adapt. These adaptations are usually designed to counteract the effects of the metabolic intervention. They usually don't negate the intervention, but they will often lessen its effects compared to the initiation phase.
3. **Maintenance**—At some point, your metabolism will become fully adapted to the intervention and your body and brain will feel more stable. You can always increase the dose or intensity of the intervention, which will then take you back to the initiation and adaptation phases again.
4. **Withdrawal**—If metabolic interventions are reduced or stopped quickly, there is usually a withdrawal reaction. Metabolism will often rebound too high or too low in these situations, which can produce withdrawal symptoms.

All these situations can happen with all the treatments I've mentioned, including medications, light therapy, changes in diet, changes in the gut microbiome, supplement use, and even psychological and social interventions.

Remember that you're looking for interventions that will improve your metabolic health over the long run. So, even if an intervention makes things worse initially, such as experiencing irritability when starting a new diet, if the maintenance phase will lead to improved metabolic health, it's worth pursuing. Obviously, it needs to be done in a way that is safe and tolerable, but getting to the maintenance phase is the goal. Likewise, other things, such as heavy alcohol use, can make things feel better in the short run (initiation phase), but can impair metabolism over the long run (maintenance phase). Stopping its use (withdrawal phase) can be particularly difficult and dangerous for heavy drinkers. It's important to keep these in mind, as starting and stopping treatments needs to be done safely.

Inpatient and Residential Treatment Programs

For some people with serious mental disorders, designing a comprehensive treatment plan might be impossible to do on their own. By definition, they have impaired brain function. They may not be able to stay on task, learn new information easily, stick to a diet, or fit all the changes into a daily schedule. It doesn't mean they can't do it, or that they won't benefit from it, but they may need help. For others, mental symptoms can sometimes threaten their safety or the safety of others and trying treatments on an outpatient basis may be dangerous. For all these people, we need to develop inpatient and residential metabolic treatment programs. These programs can customize treatment plans to people's specific needs. They can provide support not only through professionals working at the program, but through the peer support from other participants. These will be environments where everyone is working to improve their mental and metabolic health together.

Designing Your Treatment Plan

- If you have serious, dangerous, or chronic symptoms, you should work with a clinician.
- Identify contributing causes that are leading to severe metabolic impairment or threatening your safety (e.g., suicide attempts, severe substance use, living in a physically abusive environment, an

out-of-control eating disorder, severe lack of sleep, etc.). If you have any of these, you need to address them first.

- Choose one or more of the treatments outlined in this book that you think might help.
- Implement the treatment and give it at least three months to start working.
- If the treatment improves any of your symptoms, even if just a little bit, keep doing it.
- If a treatment doesn't help at all after three months, feel free to stop it.
- If a treatment helps but doesn't resolve all your symptoms, add another treatment. You are now developing your multifaceted treatment plan.
- Continue to add or try treatments until you achieve your desired results.

Your goals might change over time. Initially, you might only want to get rid of one symptom. That's fine. As life goes on, you might decide that you'd like to improve some other areas as well. Life is a journey. We all have areas of strength, but we also all have areas of weakness and vulnerability. No one is perfect. I hope that you will make the most of your life and strive to improve your health as much as possible. But I also hope that you can reach a point of gratitude for the health and strength that you do have and simply enjoy it.

Success Story: Beth—Overmedicated and Still Sick

Beth began psychiatric treatment at age nine after being diagnosed with ADHD. She was prescribed stimulants and did well in school, getting mostly

As and Bs. However, she was still impulsive and often interrupted people. As a result, she didn't have many friends, often felt criticized or rejected by others, and struggled with low self-esteem. By high school, things were much worse. She had chronic depression, was frequently suicidal, and began cutting herself with knives or razor blades. She tried antidepressants, more stimulants, mood stabilizers, anxiety medications, and even antipsychotics, but her symptoms only worsened. By the time she was in college, she had been hospitalized many times.

Beth was twenty-one when I first met her. She was diagnosed with chronic depression, panic disorder, borderline personality disorder, premenstrual dysphoric disorder, and ADHD. She was on five medications, and they clearly weren't helping. In fact, she was often lethargic and out of it. She had been in a few car accidents as a result. She had returned home during her summer break from college for more intensive psychiatric treatment. At first, I prescribed even more antipsychotic and mood-stabilizing medications, which often gave her new side effects or simply didn't work. At the same time, we began dialectical behavior therapy (DBT), which is a type of psychotherapy focused on skills that can help people better manage their emotions, suicidal impulses, and self-injury.

Beth and I both believed that medications had not helped her and, in fact, may have contributed to her problems, so we agreed to slowly taper her off them. This was extremely difficult and dangerous. Almost every time we reduced a medication, even if just by a tiny amount, she would experience an increase in symptoms—depression, suicidal impulses, or urges to cut herself. We used DBT skills to manage these symptoms and keep her safe, but we continued on with the medication taper. It took several years to get her off everything. By the time we did, things were much better. She was able to keep herself safe, had been able to hold a job, and had finished college.

The next phase of her recovery started with exercise. She began riding her bicycle outside and really got into it. She decided to work up to a charity ride. She then joined Weight Watchers to lose some weight, which helped even more. Her psychiatric symptoms were all but gone now. After many long discussions between us and her family and friends, we agreed that she no longer needed therapy or any psychiatric treatment. But that didn't stop her metabolic treatment! Beth went on to become an accomplished athlete, competing in triathlons and ultramarathons. She got married. She got a full-time job.

Today, about ten years later, Beth continues to feel great and has no psychiatric symptoms. When I ran into her father (who happens to be a physician) recently, he gave me an update on how well she was doing and said, "You know that you saved her life. You really did. I can't imagine that she would still be here today if it wasn't for you."

Beth's case illustrates the common problem of having multiple diagnoses, getting lots of treatment, and still not doing well. In fact, it appears all but certain that the medications were contributing to her problems, if not causing them. This doesn't mean that medications can't be immensely helpful for some people. I think they can be. However, for Beth, they appeared to be making matters worse. Some of the medications that she was taking are known to impair metabolism and mitochondrial function, so the brain energy theory offers an explanation for her improvement off the medications. But that wasn't enough for her full recovery. She also exercised, and lost weight, and found love, and a job, and self-respect. These all played a role in her extraordinary recovery.

Chapter 21

A New Day for Mental and Metabolic Health

The brain energy theory offers a new model of mental health. It's about much more than just brain function; it's about metabolism and mitochondria, which impact almost all aspects of human health, aging, and longevity. This new model cuts across diagnostic categories and addresses many disorders all at once. It applies not only to those diagnosed with "mental" illness, but also to those who suffer from related metabolic disorders, such as obesity, diabetes, cardiovascular disease, Alzheimer's disease, epilepsy, and chronic pain disorders. Almost all people with these disorders have at least some "mental" symptoms, and people with mental disorders are more likely to develop these related disorders. This new model offers the hope of preventing illnesses and keeping people happy, healthy, and productive for longer than is currently possible.

The theory of brain energy is a significant breakthrough that finally connects the dots to give us a clearer picture of mental illness. The science and evidence are integrated into one framework, uniting the biological, psychological, and social theories of mental illness. As we embrace the idea that mental disorders are no longer syndromes but metabolic disorders of the brain, new solutions become obvious. We need to restore brain energy by normalizing metabolism and the function of mitochondria. Once we do this, the symptoms of mental illness will begin to resolve.

The good news is that this new understanding lets us apply existing treatments more effectively and suggests novel treatment options we already have access to, including all the treatments I've shared in this book. We don't have to wait years before trying them. This doesn't mean that what we have now is enough—not all interventions will work for everyone, and we need more research and additional treatments. Finding new treatments, however, will be immensely easier now that we've connected the dots and identified the core problem. It's a problem that can now be solved with science and research, no longer some abstract mystery waiting on a miracle.

Once we begin thinking about all these disorders through the lens of metabolism and mitochondria, the possibilities for improvement are endless. We can develop diagnostic tools to assess people's metabolic health. We can develop evidence-based strategies and therapeutics to address metabolic and mitochondrial dysfunction. We can focus on better understanding the impact on mitochondria and metabolism of medications, alcohol, tobacco, recreational drugs, diets and foods, and toxins.

But we need resources to do this work. We need big changes! We need multidisciplinary healthcare teams working together to restore metabolic health in people. These teams will include physicians, nurses, psychotherapists, social workers, physical and occupational therapists, pharmacists, dietitians, personal trainers, health and wellness coaches, and many others. Health insurance will need to cover some of these costs. The biotech and pharmaceutical industries will need to rise to the challenge of developing more effective therapeutics. Government will need to be involved. We need research funding for all this work and parity for mental health services. We may also have metabolic toxins in our everyday lives that need to be regulated and/or eliminated. And of course, each and every one of us will need to do our part as well. We need self-help groups, support groups, and advocacy initiatives. We need a society that is just, fair, compassionate, peaceful, and cooperative. We need to ensure that all people have opportunities to live meaningful lives. People need to feel safe. They also benefit from feeling respected. Obviously, this is all easier said than done. In many ways, it represents a type of utopia. We all know that will take time to achieve. But we don't need to wait for utopia in order to do something.

And so, I call on *you* and ask for *your* help. To turn this hope into reality, we need a grassroots movement. Just as there were movements for HIV/AIDS and breast cancer, we desperately need a movement calling for radical changes in the way we understand and treat mental illness. Educating and informing people will take time and effort. You can help by spreading the word about the theory of brain energy. This movement needs you, your friends, and your family. I'm not asking for myself, but on behalf of the countless people who are suffering from mental illness alone and in silence; on behalf of those begging for more effective treatments and better lives; on behalf of those who are tormented on a daily basis by their mental symptoms; on behalf of those who have lost all hope; on behalf of all the people who continue to hide in their homes in shame and humiliation from the stigma of mental illness; and in memory of the countless people who couldn't hold on and are no longer with us. Let's put an end to this suffering, once and for all. Let's not waste another day.

Please visit www.brainenergy.com in order to learn more and get involved.

Acknowledgments

When I began writing *Brain Energy,* some people recommended that I make it a simple self-help book as opposed to a more rigorous science book because simple messages are easier to sell and many publishers aren't interested in science books. I want to thank my publisher, BenBella, and in particular Leah Wilson and Alexa Stevenson, for believing in me and this book and for recognizing that people are, in fact, interested in science and complexity, especially when it has the potential to transform a field and improve people's lives.

To my editor, Alexa Stevenson, thank you for your keen insights, honest feedback, and for encouraging me to include more science and evidence, not less. Your initial "healthy skepticism" of the brain energy theory pushed me to write an immensely better book.

To my literary agent, Linda Konner, thank you for your fierce representation of this work. I knew you were tough when I met you, but seeing you in action took it to a whole new level.

To the entire team at BenBella, thank you for your creativity, attention to detail, adherence to timelines, and everything else that goes into producing an amazing book.

To my early reviewers—Karen Weintraub, Anne Rauch, Julianne Torrence, Amy Yuhasz, and my brother, David Palmer—your encouragement and critiques of early drafts gave me the energy and confidence to keep writing.

Finally, to all the patients I've worked with over the past 31 years (I'm including medical school in these years), every single one of you helped me write this book, because I learned something about mental illness and the human condition from each of you. Thank you for the honor and privilege of being your psychiatrist. Thanks to the ones who were willing to try my "metabolic treatments," and especially those willing to share their stories in this book or appear publicly with me in presentations, on television, and on national radio. To the ones who never got better under my care despite trying treatment after treatment, I want to apologize on behalf of myself and my profession. It was you who taught me to question and challenge my own work and the mental health field, and to not accept the unacceptable paradigm. You inspired me to ponder endlessly about neuroscience, physiology, and human existence. You forced me to look for better answers. My greatest hope is that this book might represent the beginning of those answers.

Notes

Chapter 1

1 Saloni Dattani, Hannah Ritchie, and Max Roser. "Mental Health." OurWorldInData.org. https://ourworldindata.org/mental-health. Retrieved 10/15/2021.

2 R. C. Kessler, P. Berglund, O. Demler, R. Jin, K. R. Merikangas, and E. E. Walters. "Lifetime Prevalence and Age-of-Onset Distributions of DSM-IV Disorders in the National Comorbidity Survey Replication." *Arch Gen Psychiatry* 62(6) (2005): 593–602.

3 W. Wurm, K. Vogel, A. Holl, C. Ebner, D. Bayer, et al. "Depression-Burnout Overlap in Physicians." *PLOS ONE* 11(3): e0149913 (2016). doi: 10.1371/journal.pone.0149913.

4 Ben Wigert and Sangeeta Agrawal. "Employee Burnout, Part 1: The 5 Main Causes." Gallup. https://www.gallup.com/workplace/237059/employee-burnout-part-main-causes.aspx. Retrieved 5/28/19.

5 B. Bandelow and S. Michaelis. "Epidemiology of Anxiety Disorders in the 21st Century." *Dialogues Clin Neurosci* 17(3) (2015): 327–335. doi: 10.31887/DCNS.2015.17.3/bbandelow.

6 R. D. Goodwin, A. H. Weinberger, J. H. Kim, M. Wu, and S. Galea. "Trends in Anxiety Among Adults in the United States, 2008–2018: Rapid Increases Among Young Adults." *J Psychiatr Res.* 130 (2020): 441–446. doi: 10.1016/j.jpsychires.2020.08.014.

7 SAMHSA. "National Survey on Drug Use and Health: Comparison of 2008–2009 and 2016–2017 Population Percentages (50 States and the District of Columbia)." Substance Abuse and Mental Health Services Administration, US Department of Health and Human Services. https://www.samhsa.gov/data/sites/default/files/cbhsq-reports/NSDUHsaeTrendTabs2017/NSDUHsaeLongTermCHG2017.pdf. Retrieved 2/18/22.

8 CDC. "Data & Statistics on Autism Spectrum Disorder." Centers for Disease Control and Prevention, US Department of Health and Human Services. https://www.cdc.gov/ncbddd/autism/data.html. Retrieved 5/27/19.

9 S. H. Yutzy, C. R. Woofter, C. C. Abbott, I. M. Melhem, and B. S. Parish. "The Increasing Frequency of Mania and Bipolar Disorder: Causes and Potential Negative Impacts." *J Nerv Ment Dis.* 200(5) (2012): 380–387. doi: 10.1097/NMD.0b013e3182531f17.

10 M. É. Czeisler, R. I. Lane, E. Petrosky, et al. "Mental Health, Substance Use, and Suicidal Ideation During the COVID-19 Pandemic—United States, June 24–30, 2020." *MMWR Morb Mortal Wkly Rep* 69 (2020): 1049–1057. doi: 10.15585/mmwr.mm6932a1external icon.

11 The Lancet Global Health. "Mental Health Matters." *Lancet Glob Health* 8(11) (November 2020): e1352.

12 Global Burden of Disease Collaborative Network. "Global Burden of Disease Study 2015 (GBD 2015) Life Expectancy, All-Cause and Cause-Specific Mortality 1980–2015." Seattle, United States: Institute for Health Metrics and Evaluation (IHME), 2016.

13 US Department of Housing and Urban Development. "The 2010 Annual Homeless Assessment Report to Congress." US Department of Housing and Urban Development. https://www.huduser.gov/portal/sites/default/files/pdf/2010HomelessAssessmentReport.pdf. Retrieved 7/24/21.

14 Doris J. James and Lauren E. Glaze. "Mental Health Problems of Prison and Jail Inmates." Bureau of Justice Statistics, US Dept. of Justice (September 2006). https://bjs.ojp.gov/library/publications/mental-health-problems-prison-and-jail-inmates. Retrieved 7/24/21.

15 National Institute of Mental Health. "Major Depression." National Institute of Mental Health, US Dept. of Health and Human Services. https://www.nimh.nih.gov/health/statistics/major-depression#:~:text=all%20U.S.%20adults.-,Treatment%20of%20Major%20Depressive%20Episode%20Among%20Adults,treatment%20in%20the%20past%20year. Retrieved 2/18/2022.

16 L. L. Judd, H. S. Akiskal, J. D. Maser, et al. "A Prospective 12-Year Study of Subsyndromal and Syndromal Depressive Symptoms in Unipolar Major Depressive Disorders." *Arch Gen Psychiatry.* 55(8) (1998): 694–700. doi: 10.1001/archpsyc.55.8.694.

17 Sidney Zisook, Gary R. Johnson, Ilanit Tal, Paul Hicks, Peijun Chen, Lori Davis, Michael Thase, Yinjun Zhao, Julia Vertrees, and Somaia Mohamed. "General Predictors and Moderators of Depression Remission: A VAST-D Report." *Am. J Psychiatry* 176(5) (May 1, 2019): 348–357. doi: 10.1176/appi.ajp.2018.18091079.

18 Diego Novick, Josep Maria Haro, David Suarez, Eduard Vieta, and Dieter Naber. "Recovery in the Outpatient Setting: 36-Month Results from the Schizophrenia Outpatients Health Outcomes (SOHO) Study." *Schizophr Res* 108(1) (2009): 223–230. doi: 10.1016/j.schres.2008.11.007.

19 Adam Rogers. "Star Neuroscientist Tom Insel Leaves the Google-Spawned Verily for . . . a Startup?" *Wired.* May 11, 2017. https://www.wired.com/2017/05/star-neuroscientist-tom-insel-leaves-google-spawned-verily-startup/#:~:text=%E2%80%9CI%20spent%2013%20years%20at,we%20moved%20the%20needle%20in.

Chapter 2

1 G. L. Engel. "The Need for a New Medical Model: A Challenge for Biomedicine." *Science* 196(4286) (1977): 129–136. doi: 10.1126/science.847460.

2 M. B. Howren, D. M. Lamkin, and J. Suls. "Associations of Depression with C-Reactive Protein, IL-1, and IL-6: A Meta-Analysis." *Psychosom Med.* 71(2) (February 2009): 171–186. doi: 10.1097/PSY.0b013e3181907c1b.

3 E. Setiawan, S. Attwells, A. A. Wilson, R. Mizrahi, P. M. Rusjan, L. Miler, C. Xu, S. Sharma, S. Kish, S. Houle, and J. H. Meyer. "Association of Translocator Protein Total Distribution Volume with Duration of Untreated Major Depressive Disorder: A Cross-Sectional Study." *Lancet Psychiatry* 5(4) (April 2018): 339–347. doi: 10.1016/S2215-0366(18)30048-8.

4 C. Zhuo, G. Li, X. Lin, et al. "The Rise and Fall of MRI Studies in Major Depressive Disorder." *Transl Psychiatry* 9(335) (2019). doi.org/10.1038/s41398-019-0680-6.

5 A. L. Komaroff. "The Microbiome and Risk for Obesity and Diabetes." *JAMA* 317(4) (2017): 355–356. doi: 10.1001/jama.2016.20099; K. E. Bouter, D. H. van Raalte, A. K. Groen, et al. "Role of the Gut Microbiome in the Pathogenesis of Obesity and Obesity-Related Metabolic Dysfunction." *Gastroenterology* 152(7) (May 2017): 1671–1678. doi: 10.1053/j.gastro.2016.12.048; E. A. Mayer, K. Tillisch, and A. Gupta. "Gut/Brain Axis and the Microbiota." *J Clin Invest* 125(3) (2015): 926–938. doi: 10.1172/JCI76304.

6 J. A. Foster and K. A. McVey Neufeld. "Gut-Brain Axis: How the Microbiome Influences Anxiety and Depression." *Trends Neurosci* 36(5) (May 2013): 305–312. doi: 10.1016/j.tins.2013.01.005.

Chapter 3

1 American Psychiatric Association. *Diagnostic and Statistical Manual of Mental Disorders: DSM-IV-TR.* 4th ed. Arlington, VA: American Psychiatric Association, 2000: 356.

2 E. Corruble, B. Falissard, and P. Gorwood. "Is DSM-IV Bereavement Exclusion for Major Depression Relevant to Treatment Response? A Case-Control, Prospective Study." *J Clin Psychiatry* 72(7) (July 2011): 898–902. doi: 10.4088/JCP.09m05681blu.

3 Alan F. Schatzberg. "Scientific Issues Relevant to Improving the Diagnosis, Risk Assessment, and Treatment of Major Depression." *Am J Psychiatry* 176(5) (2019): 342–47. doi: 10.1176/appi.ajp.2019.19030273.

4 M. K. Jha, A. Minhajuddin, C. South, A. J. Rush, and M. H. Trivedi. "Irritability and Its Clinical Utility in Major Depressive Disorder: Prediction of Individual-Level Acute-Phase Outcomes Using Early Changes in Irritability and Depression Severity." *Am J Psychiatry* 176(5) (May 1, 2019): 358–366. doi: 10.1176/appi.ajp.2018.18030355. Epub Mar 29, 2019. PMID: 30922100.

5 Maurice M. Ohayon and Alan F. Schatzberg. "Chronic Pain and Major Depressive Disorder in the General Population." *J Psychiatr Res* 44(7) (2010): 454–61. doi: 10.1016/j.jpsychires.2009.10.013.

6 R. C. Kessler, W. T. Chiu, O. Demler, and E. E. Walters. "Prevalence, Severity, and Comorbidity of 12-Month DSM-IV Disorders in the National Comorbidity Survey Replication." *Arch Gen Psychiatry* 62(6) (2005): 617–627. doi: 10.1001/archpsyc.62.6.617.

7 R. C. Kessler, P. Berglund, O. Demler, et al. "The Epidemiology of Major Depressive Disorder: Results from the National Comorbidity Survey Replication (NCS-R)." *JAMA* 289 (2003): 3095–3105. doi: 10.1001/jama.289.23.3095; B. W. Penninx, D. S. Pine, E. A. Holmes, and A. Reif. "Anxiety Disorders." *Lancet* 397(10277) (2021): 914–927.

8 M. Olfson, S. C. Marcus, and J. G. Wan. "Treatment Patterns for Schizoaffective Disorder and Schizophrenia Among Medicaid Patients." *Psychiatr Serv* 60 (2009): 210–216. doi: 10.1176/ps.2009.60.2.210.

9 Seth Himelhoch, Eric Slade, Julie Kreyenbuhl, Deborah Medoff, Clayton Brown, and Lisa Dixon. "Antidepressant Prescribing Patterns Among VA Patients with Schizophrenia." *Schizophr Res* 136(1) (2012): 32–35. doi: 10.1016/j.schres.2012.01.008.

10 P. D. Harvey, R. K. Heaton, W. T. Carpenter Jr., M. F. Green, J. M. Gold, and M. Schoenbaum. "Functional Impairment in People with Schizophrenia: Focus on Employability and Eligibility for Disability Compensation." *Schizophr Res* 140(1–3) (2012): 1–8. doi: 10.1016/j.schres.2012.03.025.

11 L. L. Judd, H. S. Akiskal, P. J. Schettler, et al. "The Long-Term Natural History of the Weekly Symptomatic Status of Bipolar I Disorder." *Arch Gen Psychiatry* 59(6) (June 2002): 530–537. doi: 10.1001/archpsyc.59.6.530.

12 "Biomarkers Outperform Symptoms in Parsing Psychosis Subgroups." National Institutes of Health. December 8, 2015. https://www.nih.gov/news-events/news-releases/biomarkers-outperform-symptoms-parsing-psychosis-subgroups.

13 Maurice M. Ohayon and Alan F. Schatzberg. "Prevalence of Depressive Episodes with Psychotic Features in the General Population." *Am J Psychiatry* 159(11) (2002): 1855–1861. doi: 10.1176/appi.ajp.159.11.1855.

14 B. Bandelow and S. Michaelis. "Epidemiology of Anxiety Disorders in the 21st Century." *Dialogues Clin Neurosci* 17(3) (2015): 327–335. doi: 10.31887/DCNS.2015.17.3/bbandelow.

15 O. Plana-Ripoll, C. B. Pedersen, Y. Holtz, et al. "Exploring Comorbidity Within Mental Disorders Among a Danish National Population." *JAMA Psychiatry* 76(3) (2019): 259–270. doi: 10.1001/jamapsychiatry.2018.3658.

16 R. C. Kessler, W. T. Chiu, O. Demler, and E. E. Walters. "Prevalence, Severity, and Comorbidity of 12-Month DSM-IV Disorders in the National Comorbidity Survey Replication." *Arch Gen Psychiatry* 62(6) (2005): 617–627. doi: 10.1001/archpsyc.62.6.617.

17 M. C. Lai, C. Kassee, R. Besney, S. Bonato, L. Hull, W. Mandy, P. Szatmari, and S. H. Ameis. "Prevalence of Co-occurring Mental Health Diagnoses in the Autism Population: A Systematic Review and Meta-Analysis." *Lancet Psychiatry* 6(10) (October 2019): 819–829. doi: 10.1016/S2215-0366(19)30289-5.

18 O. Plana-Ripoll, C. B. Pedersen, Y. Holtz, et al. "Exploring Comorbidity Within Mental Disorders Among a Danish National Population." *JAMA Psychiatry* 76(3) (2019): 259–270. doi: 10.1001/jamapsychiatry.2018.3658.

19 National Institute of Mental Health. "Eating Disorders." National Institute of Mental Health, US Dept. of Health and Human Services. https://www.nimh.nih.gov/health/statistics/eating-disorders.shtml. Retrieved 7/24/21.

20 K. R. Merikangas, J. P. He, M. Burstein, S. A. Swanson, S. Avenevoli, L. Cui, C. Benjet, K. Georgiades, and J. Swendsen. "Lifetime Prevalence of Mental Disorders in U.S. Adolescents: Results from the National

Comorbidity Survey Replication—Adolescent Supplement (NCS-A)." *J Am Acad Child Adolesc Psychiatry* 49(10) (October 2010): 980–989. http://www.ncbi.nlm.nih.gov/pubmed/20855043/.

21 O. Plana-Ripoll, C. B. Pedersen, Y. Holtz, et al. "Exploring Comorbidity Within Mental Disorders Among a Danish National Population." *JAMA Psychiatry* 76(3) (2019): 259–270. doi: 10.1001/jamapsychiatry.2018.3658.

22 B. B. Lahey, B. Applegate, J. K. Hakes, D. H. Zald, A. R. Hariri, and P. J. Rathouz. "Is There a General Factor of Prevalent Psychopathology During Adulthood?" *J Abnorm Psychol* 121(4) (2012): 971–977. doi: 10.1037/a0028355.

23 Avshalom Caspi and Terrie E. Moffitt. "All for One and One for All: Mental Disorders in One Dimension." *Am J Psychiatry* 175(9) (2018): 831–44. doi: 10.1176/appi.ajp.2018.17121383.

24 E. Pettersson, H. Larsson, and P. Lichtenstein. "Common Psychiatric Disorders Share the Same Genetic Origin: A Multivariate Sibling Study of the Swedish Population." *Mol Psychiatry* 21 (2016): 717–721. doi: 10.1038/mp.2015.116.

25 A. Caspi, R. M. Houts, A. Ambler, et al. "Longitudinal Assessment of Mental Health Disorders and Comorbidities Across 4 Decades Among Participants in the Dunedin Birth Cohort Study." *JAMA Netw Open* 3(4) (2020): e203221. doi: 10.1001/jamanetworkopen.2020.3221.

Chapter 4

1 A. P. Rajkumar, H. T. Horsdal, T. Wimberley, et al. "Endogenous and Antipsychotic-Related Risks for Diabetes Mellitus in Young People with Schizophrenia: A Danish Population-Based Cohort Study." *Am J Psychiatry* 174 (2017): 686–694. doi: 10.1176/appi.ajp.2016.16040442.

2 B. Mezuk, W. W. Eaton, S. Albrecht, and S. H. Golden. "Depression and Type 2 Diabetes over the Lifespan: A Meta-Analysis." *Diabetes Care* 31 (2008): 2383–2390. doi: 10.2337/dc08-0985.

3 K. Semenkovich, M. E. Brown, D. M. Svrakic, et al. "Depression and Diabetes." *Drugs* 75(6) (2015): 577. doi: 10.1007/s40265-015-0347-4.

4 M. E. Robinson, M. Simard, I. Larocque, J. Shah, M. Nakhla, and E. Rahme. "Risk of Psychiatric Disorders and Suicide Attempts in Emerging Adults with Diabetes." *Diabetes Care* 43(2) (2020): 484–486. doi: 10.2337/dc19-1487.

5 Martin Strassnig, Roman Kotov, Danielle Cornaccio, Laura Fochtmann, Philip D. Harvey, and Evelyn J. Bromet. "Twenty-Year Progression of Body Mass Index in a County-Wide Cohort of People with Schizophrenia and Bipolar Disorder Identified at Their First Episode of Psychosis." *Bipolar Disord* 19(5) (2017): 336–343. doi: 10.1111/bdi.12505.

6 L. Mische Lawson and L. Foster. "Sensory Patterns, Obesity, and Physical Activity Participation of Children with Autism Spectrum Disorder." *Am J Occup Ther* 70(5) (2016): 7005180070pl-7005180070p8. doi: 10.5014/ajot.2016.021535.

7 M. Afzal, N. Siddiqi, B. Ahmad, N. Afsheen, F. Aslam, A. Ali, R. Ayesha, M. Bryant, R. Holt, H. Khalid, K. Ishaq, K. N. Koly, S. Rajan, J. Saba, N. Tirbhowan, and G. A. Zavala. "Prevalence of Overweight and Obesity in People with Severe Mental Illness: Systematic Review and Meta-Analysis." *Front Endocrinol* (Lausanne) 12 (2021): 769309. doi: 10.3389/fendo.2021.769309.

8 M. Shaw, P. Hodgkins, H. Caci, S. Young, J. Kahle, A. G. Woods, and L. E. Arnold. "A Systematic Review and Analysis of Long-Term Outcomes in Attention Deficit Hyperactivity Disorder: Effects of Treatment and Non-Treatment." *BMC Med* 10 (2012): 99. doi: 10.1186/1741-7015-10-99.

9 B. I. Perry, J. Stochl, R. Upthegrove, et al. "Longitudinal Trends in Childhood Insulin Levels and Body Mass Index and Associations with Risks of Psychosis and Depression in Young Adults." *JAMA Psychiatry*. Published online January 13, 2021. doi: 10.1001/jamapsychiatry.2020.4180.

10 V. C. Chen, Y. C. Liu, S. H. Chao, et al. "Brain Structural Networks and Connectomes: The Brain-Obesity Interface and its Impact on Mental Health." *Neuropsychiatr Dis Treat* 14 (November 26, 2018): 3199–3208. doi:10.2147/NDT.S180569; K. Thomas, F. Beyer, G. Lewe, et al. "Higher Body Mass Index Is Linked to Altered Hypothalamic Microstructure." *Sci Rep* 9(1) (2019): 17373. doi: 10.1038/s41598-019-53578-4.

11 M. Åström, R. Adolfsson, and K. Asplund. "Major Depression in Stroke Patients: A 3-year Longitudinal Study." *Stroke* 24(7) (1993): 976–982. doi: 10.1161/01.STR.24.7.976.

12 Heather S. Lett, James A. Blumenthal, Michael A. Babyak, Andrew Sherwood, Timothy Strauman, Clive Robins, and Mark F. Newman. "Depression as a Risk Factor for Coronary Artery Disease: Evidence, Mechanisms, and Treatment." *Psychosom Med* 66(3) (2004):305–15. doi: 10.1097/01.psy.0000126207.43307.c0.

13 Z. Fan, Y. Wu, J. Shen, T. Ji, and R. Zhan. "Schizophrenia and the Risk of Cardiovascular Diseases: A Meta-Analysis of Thirteen Cohort Studies." *J Psychiatr Res* 47(11) (2013): 1549–1556. doi: 10.1016/j.jpsychires.2013.07.011.

14 Lindsey Rosman, Jason J. Sico, Rachel Lampert, Allison E. Gaffey, Christine M. Ramsey, James Dziura, Philip W. Chui, et al. "Post-traumatic Stress Disorder and Risk for Stroke in Young and Middle-Aged Adults." *Stroke* 50(11) (2019): STROKEAHA.119.026854. doi: 10.1161/STROKEAHA.119.026854.

15 C. W. Colton and R. W. Manderscheid. "Congruencies in Increased Mortality Rates, Years of Potential Life Lost, and Causes of Death Among Public Mental Health Clients in Eight States." *Prev Chronic Dis* [serial online] (April 2006 [*date cited*]). Available from: http://www.cdc.gov/pcd/issues/2006/apr/05_0180.htm.

16 Oleguer Plana-Ripoll, et al. "A Comprehensive Analysis of Mortality-related Health Metrics Associated With Mental Disorders: A Nationwide, Register-based Cohort Study." *Lancet* 394(10211) (2019): 1827–1835. doi: 10.1016/S0140-6736(19)32316-5.

17 S. E. Bojesen. "Telomeres and Human Health." *J Intern Med* 274(5) (2013): 399–413. doi: 10.1111/joim.12083.

18 Alzheimer's Association. "2022 Alzheimer's Disease Facts and Figures." *Alzheimers Dement* 18(4) (2022): 700–789. doi: 10.1002/alz.12638.

19 R. L. Ownby, E. Crocco, A. Acevedo, V. John, and D. Loewenstein. "Depression and Risk for Alzheimer Disease: Systematic Review, Meta-Analysis, and Meta-Regression Analysis." *Arch Gen Psychiatry* 63(5) (2006): 530–538. doi: 10.1001/archpsyc.63.5.530.

20 T. S. Stroup, M. Olfson, C. Huang, et al. "Age-Specific Prevalence and Incidence of Dementia Diagnoses Among Older US Adults with Schizophrenia." *JAMA Psychiatry* 78(6) (2021): 632–641. doi: 10.1001/jamapsychiatry.2021.0042.

21 M. Steinberg, H. Shao, P. Zandi, et al. "Point and 5-Year Period Prevalence of Neuropsychiatric Symptoms in Dementia: The Cache County Study." *Int J Geriatr Psychiatry* 23(2) (2008): 170–177. doi: 10.1002/gps.1858.

22 P. S. Murray, S. Kumar, M. A. Demichele-Sweet, R. A. Sweet. "Psychosis in Alzheimer's Disease." *Biol Psychiatry* 75(7) (2014): 542–552. doi: 10.1016/j.biopsych.2013.08.020.

23 Colin Reilly, Patricia Atkinson, Krishna B. Das, Richard F. M. C. Chin, Sarah E. Aylett, Victoria Burch, Christopher Gillberg, Rod C. Scott, and Brian G. R. Neville. "Neurobehavioral Comorbidities in Children with Active Epilepsy: A Population-Based Study." *Pediatrics* 133(6) (2014): e1586. doi: 10.1542/peds.2013-3787.

24 A. M. Kanner. "Anxiety Disorders in Epilepsy: The Forgotten Psychiatric Comorbidity." *Epilepsy Curr* 11(3) (2011): 90–91. doi: 10.5698/1535-7511-11.3.90.

25 M. F. Mendez, J. L. Cummings, and D. F. Benson. "Depression in Epilepsy: Significance and Phenomenology." *Arch Neurol* 43(8) (1986): 766–770. doi: 10.1001/archneur.1986.00520080014012.

26 C. E. Elger, S. A. Johnston, and C. Hoppe. "Diagnosing and Treating Depression in Epilepsy." *Seizure* 44(1) (2017): 184–193. doi: 10.1016/j.seizure.2016.10.018.

27 Alan B. Ettinger, Michael L. Reed, Joseph F. Goldberg, and Robert M.A. Hirschfeld. "Prevalence of Bipolar Symptoms in Epilepsy vs. Other Chronic Health Disorders." *Neurology* 65(4) (2005): 535. doi: 10.1212/01.wnl.0000172917.70752.05; Mario F. Mendez, Rosario Grau, Robert C. Doss, and Jody L. Taylor. "Schizophrenia in Epilepsy: Seizure and Psychosis Variables." *Neurology* 43(6) (1993): 1073-7. doi: 10.1212/wnl.43.6.1073.

28 S. S. Jeste and R. Tuchman. "Autism Spectrum Disorder and Epilepsy: Two Sides of the Same Coin?" *J Child Neurol* 30(14) (2015): 1963–1971. doi: 10.1177/0883073815601501.

29 E. H. Lee, Y. S. Choi, H. S. Yoon, and G. H. Bahn. "Clinical Impact of Epileptiform Discharge in Children with Attention-Deficit/Hyperactivity Disorder (ADHD)." *J Child Neurol* 31(5) (2016): 584–588. doi: 10.1177/0883073815604223.

30 D. C. Hesdorffer, P. Ludvigsson, E. Olafsson, G. Gudmundsson, O. Kjartansson, and W. A. Hauser. "ADHD as a Risk Factor for Incident Unprovoked Seizures and Epilepsy in Children." *Arch Gen Psychiatry* 61(7) (2004): 731–736. doi: 10.1001/archpsyc.61.7.731.

31 D. C. Hesdorffer, W. A. Hauser, and J. F. Annegers. "Major Depression Is a Risk Factor for Seizures in Older Adults." *Ann Neurol* 47(2) (2001): 246–249. doi: 10.1002/1531-8249(200002)47:2%3C246::AID-ANA17%3E3.0.CO;2-E.

32 G. E. Dafoulas, K. A. Toulis, D. Mccorry, et al. "Type 1 Diabetes Mellitus and Risk of Incident Epilepsy: A Population-Based, Open-Cohort Study." *Diabetologia* 60(2) (2017): 258–261. doi: 10.1007/s00125-016-4142-x.

33 I. C. Chou, C. H. Wang, W. D. Lin, F. J. Tsai, C. C. Lin, and C. H. Kao. "Risk of Epilepsy in Type 1 Diabetes Mellitus: A Population-Based Cohort Study." *Diabetologia* 59 (2016): 1196–1203. doi: 10.1007/s00125-016-3929-0.

34 M. Baviera, M. C. Roncaglioni, M. Tettamanti, et al. "Diabetes Mellitus: A Risk Factor for Seizures in the Elderly—A Population-Based Study." *Acta Diabetol* 54 (2017): 863. doi: 10.1007/s00592-017-1011-0.

35 S. Gao, J. Juhaeri, and W. S. Dai. "The Incidence Rate of Seizures in Relation to BMI in UK Adults." *Obesity* 16 (2008): 2126–2132. doi: 10.1038/oby.2008.310.

36 N. Razaz, K. Tedroff, E. Villamor, and S. Cnattingius. "Maternal Body Mass Index in Early Pregnancy and Risk of Epilepsy in Offspring." *JAMA Neurol* 74(6) (2017): 668–676. doi: 10.1001/jamaneurol.2016.6130.

Chapter 5

1 Albert Einstein and Leopold Infeld. *The Evolution of Physics*. Edited by C. P. Snow. (Cambridge: Cambridge University Press, 1938).

2 F. A. Azevedo, L. R. Carvalho, L. T. Grinberg, J. M. Farfel, R. E. Ferretti, R. E. Leite, W. J. Filho, R. Lent, and S. Herculano-Houzel. "Equal Numbers of Neuronal and Nonneuronal Cells Make the Human Brain an Isometrically Scaled-Up Primate Brain." *J Comp Neurol* 513 (2009): 532–541. doi: 10.1002/cne.21974.

Chapter 6

1 J. D. Gray, T. G. Rubin, R. G. Hunter, and B. S. McEwen. "Hippocampal Gene Expression Changes Underlying Stress Sensitization and Recovery." *Mol Psychiatry* 19(11) (2014): 1171–1178. doi: 10.1038/mp.2013.175.

2 K. Hughes, M. A. Bellis, K. A. Hardcastle, D. Sethi, A. Butchart, C. Mikton, L. Jones, and M. P. Dunne. "The Effect of Multiple Adverse Childhood Experiences on Health: A Systematic Review and Meta-Analysis." *Lancet Public Health* 2(8) (August 2017): e356–e366. doi: 10.1016/S2468-2667(17)30118-4.

3 D. W. Brown, R. F. Anda, H. Tiemeier, V. J. Felitti, V. J. Edwards, J. B. Croft, and W. H. Giles. "Adverse Childhood Experiences and the Risk of Premature Mortality." *Am J Prev Med* 37(5) (2009): 389–396. doi: 10.1016/j.amepre.2009.06.021.

4 M. Sato, E. Ueda, A. Konno, H. Hirai H, Y. Kurauchi, A. Hisatsune, H. Katsuki, and T. Seki. "Glucocorticoids Negatively Regulates Chaperone Mediated Autophagy and Microautophagy." *Biochem Biophys Res Commun* 528(1) (July 12, 2020): 199–205. doi: 10.1016/j.bbrc.2020.04.132.

5 N. Mizushima and B. Levine. "Autophagy in Human Diseases." *N Engl J Med* 383(16) (October 15, 2020): 1564–1576. doi: 10.1056/NEJMra2022774; Tamara Bar-Yosef, Odeya Damri, and Galila Agam. "Dual Role of Autophagy in Diseases of the Central Nervous System." *Front Cellular Neurosci* 13 (2019): 196. doi: 10.3389/fncel.2019.00196.

6 Daniel J. Klionsky, Giulia Petroni, Ravi K. Amaravadi, Eric H. Baehrecke, Andrea Ballabio, Patricia Boya, José Manuel Bravo-San Pedro, et al. "Autophagy in Major Human Diseases." *The EMBO Journal* 40(19) (2021): e108863. doi: 10.15252/embj.2021108863.

7 J. R. Buchan and R. Parker. "Eukaryotic Stress Granules: The Ins and Outs of Translation." *Mol Cell* 36(6) (2009): 932–941. doi: 10.1016/j.molcel.2009.11.020.

8 J. M. Silva, S. Rodrigues, B. Sampaio-Marques, et al. "Dysregulation of Autophagy and Stress Granule-Related Proteins in Stress-Driven Tau Pathology." *Cell Death Differ* 26 (2019): 1411–1427. doi: 10.1038/s41418-018-0217-1.

9 E. S. Epel, E. H. Blackburn, J. Lin, F. S. Dhabhar, N. E. Adler, J. D. Morrow, and R. M. Cawthon. "Accelerated Telomere Shortening in Response to Life Stress." *Proc Natl Acad Sci USA* 101(49) (2004): 17312–17315. doi: 10.1073/pnas.0407162101.

10 B. L. Miller. "Science Denial and COVID Conspiracy Theories: Potential Neurological Mechanisms and Possible Responses." *JAMA* 324(22) (2020): 2255–2256. doi: 10.1001/jama.2020.21332.

11 K. Maijer, M. Hayward, C. Fernyhough, et al. "Hallucinations in Children and Adolescents: An Updated Review and Practical Recommendations for Clinicians." *Schizophr Bull* 45(45 Suppl 1) (2019): S5–S23. doi: 10.1093/schbul/sby119.

12 M. Ohayon, R. Priest, M. Caulet, and C. Guilleminault. "Hypnagogic and Hypnopompic Hallucinations: Pathological Phenomena?" *British Journal of Psychiatry* 169(4) (1996): 459–467. doi: 10.1192/bjp.169.4.459.

13 C. Zhuo, G. Li, X. Lin, et al. "The Rise and Fall of MRI Studies in Major Depressive Disorder." *Transl Psychiatry* 9(1) (2019): 335. doi: 10.1038/s41398-019-0680-6.

14 B. O. Rothbaum, E. B. Foa, D. S. Riggs, T. Murdock, and W. Walsh. "A Prospective Examination of Post-Traumatic Stress Disorder in Rape Victims." *J. Trauma Stress* 5 (1992): 455–475. doi: 10.1002/jts.2490050309.

Chapter 7

1 Nick Lane. *Power, Sex, Suicide: Mitochondria and the Meaning of Life* (Oxford: Oxford University Press, 2005).

2 Siv G. E. Andersson, Alireza Zomorodipour, Jan O. Andersson, Thomas Sicheritz-Pontén, U. Cecilia M. Alsmark, Raf M. Podowski, A. Kristina Näslund, Ann-Sofie Eriksson, Herbert H. Winkler, and Charles G. Kurland. "The Genome Sequence of *Rickettsia Prowazekii* and the Origin of Mitochondria." *Nature* 396(6707) (1998): 133–40. doi: 10.1038/24094.

3 Lane. *Power, Sex, Suicide.*

4 Lane. *Power, Sex, Suicide.*

5 X. H. Zhu, H. Qiao, F. Du, et al. "Quantitative Imaging of Energy Expenditure in Human Brain." *Neuroimage* 60(4) (2012): 2107–2117. doi: 10.1016/j.neuroimage.2012.02.013.

6 R. L. Frederick and J. M. Shaw. "Moving Mitochondria: Establishing Distribution of an Essential Organelle." *Traffic* 8(12) (2007): 1668–1675. doi: 10.1111/j.1600-0854.2007.00644.x.

7 D. Safiulina and A. Kaasik. "Energetic and Dynamic: How Mitochondria Meet Neuronal Energy Demands." *PLoS Biol* 11(12) (2013): e1001755. doi: 10.1371/journal.pbio.1001755.

8 R. L. Frederick and J. M. Shaw. "Moving Mitochondria: Establishing Distribution of an Essential Organelle." *Traffic* 8(12) (2007): 1668–1675. doi: 10.1111/j.1600-0854.2007.00644.x.

9 R. Rizzuto, P. Bernardi, and T. Pozzan. "Mitochondria as All-Round Players of the Calcium Game." *J Physiol* 529 Pt 1(Pt 1) (2000): 37–47. doi: 10.1111/j.1469-7793.2000.00037.x.

10 Z. Gong, E. Tas, and R. Muzumdar. "Humanin and Age-Related Diseases: A New Link?" *Front Endocrinol* (Lausanne) 5 (2014): 210. doi: 10.3389/fendo.2014.00210.

11 S. Kim, J. Xiao, J. Wan, P. Cohen, and K. Yen. "Mitochondrially Derived Peptides as Novel Regulators of Metabolism." *J Physiol* 595 (2017): 6613–6621. doi: 10.1113/JP274472.

12 L. Guo, J. Tian, and H. Du. "Mitochondrial Dysfunction and Synaptic Transmission Failure in Alzheimer's Disease." *J Alzheimers Dis* 57(4) (2017): 1071–1086. doi: 10.3233/JAD-160702.

13 Sergej L. Mironov and Natalya Symonchuk. "ER Vesicles and Mitochondria Move and Communicate at Synapses." *Journal of Cell Science* 119(23) (2006): 4926. doi: 10.1242/jcs.03254.

14 Sanford L. Palay. "Synapses in the Central Nervous System." *J Biophys and Biochem Cytol* 2(4) (1956): 193. doi: 10.1083/jcb.2.4.193.

15 Alexandros K. Kanellopoulos, Vittoria Mariano, Marco Spinazzi, Young Jae Woo, Colin McLean, Ulrike Pech, Ka Wan Li, et al. "Aralar Sequesters GABA into Hyperactive Mitochondria, Causing Social Behavior Deficits." *Cell* 180(6) (2020): 1178–1197.e20. doi: 10.1016/j.cell.2020.02.044.

16 A. West, G. Shadel, and S. Ghosh. "Mitochondria in Innate Immune Responses." *Nat Rev Immunol* 11(6) (2011): 389–402. doi: 10.1038/nri2975.

17 A. Meyer, G. Laverny, L. Bernardi, et al. "Mitochondria: An Organelle of Bacterial Origin Controlling Inflammation." *Front Immunol* 9 (2018): 536. doi: 10.3389/fimmu.2018.00536.

18 Sebastian Willenborg, David E. Sanin, Alexander Jais, Xiaolei Ding, Thomas Ulas, Julian Nüchel, Milica Popović, et al. "Mitochondrial Metabolism Coordinates Stage-Specific Repair Processes in Macrophages During Wound Healing." *Cell Metab* 33(12) (2021): 2398–2414. doi: 10.1016/j.cmet.2021.10.004.

19 L. Galluzzi, T. Yamazaki, and G. Kroemer. "Linking Cellular Stress Responses to Systemic Homeostasis." *Nat Rev Mol Cell Biol* 19(11) (2018): 731–745. doi: 10.1038/s41580-018-0068-0.

20 M. Picard, M. J. McManus, J. D. Gray, et al. "Mitochondrial Functions Modulate Neuroendocrine, Metabolic, Inflammatory, and Transcriptional Responses to Acute Psychological Stress." *Proc Natl Acad Sci USA* 112(48) (2015): E6614–E6623. doi: 10.1073/pnas.1515733112.

21 M. P. Murphy. "How Mitochondria Produce Reactive Oxygen Species." *Biochem J* 417(1) (2009): 1–13. doi: 10.1042/BJ20081386.

22 Edward T. Chouchani, Lawrence Kazak, Mark P. Jedrychowski, Gina Z. Lu, Brian K. Erickson, John Szpyt, Kerry A. Pierce, et al. "Mitochondrial ROS Regulate Thermogenic Energy Expenditure and Sulfenylation of UCP1." *Nature* 532(7597) (2016): 112. doi: 10.1038/nature17399.

23 S. Reuter, S. C. Gupta, M. M. Chaturvedi, and B. B. Aggarwal. "Oxidative Stress, Inflammation, and Cancer: How Are They Linked?" *Free Radic Biol Med* 49(11) (2010): 1603–1616. doi: 10.1016/j.freeradbiomed.2010.09.006.

24 A. Y. Andreyev, Y. E. Kushnareva, and A. A. Starkov. "Mitochondrial Metabolism of Reactive Oxygen Species." *Biochemistry* (Mosc.) 70(2) (2005): 200–214. doi: 10.1007/s10541-005-0102-7.

25 M. Schneeberger, M. O. Dietrich, D. Sebastián, et al. "Mitofusin 2 in POMC Neurons Connects ER Stress with Leptin Resistance and Energy Imbalance." *Cell* 155(1) (2013): 172–187. doi: 10.1016/j.cell.2013.09.003; M. O. Dietrich, Z. W. Liu, and T. L. Horvath. "Mitochondrial Dynamics Controlled by Mitofusins Regulate Agrp Neuronal Activity and Diet-Induced Obesity." *Cell* 155(1) (2013): 188–199. doi: 10.1016/j.cell.2013.09.004.

26 Petras P. Dzeja, Ryan Bortolon, Carmen Perez-Terzic, Ekshon L. Holmuhamedov, and Andre Terzic. "Energetic Communication Between Mitochondria and Nucleus Directed by Catalyzed Phosphotransfer." *Proc Natl Acad Sci USA* 99(15) (2002): 10156. doi: 10.1073/pnas.152259999.

27 E.A. Schroeder, N. Raimundo, and G. S. Shadel. "Epigenetic Silencing Mediates Mitochondria Stress-Induced Longevity." *Cell Metab* 17(6) (2013): 954–964. doi: 10.1016/j.cmet.2013.04.003.

28 M. D. Cardamone, B. Tanasa, C. T. Cederquist, et al. "Mitochondrial Retrograde Signaling in Mammals Is Mediated by the Transcriptional Cofactor GPS2 via Direct Mitochondria-to-Nucleus Translocation." *Mol Cell* 69(5) (2018): 757–772.e7. doi: 10.1016/j.molcel.2018.01.037.

29 K. H. Kim, J. M. Son, B. A. Benayoun, and C. Lee. "The Mitochondrial-Encoded Peptide MOTS-c Translocates to the Nucleus to Regulate Nuclear Gene Expression in Response to Metabolic Stress." *Cell Metab* 28(3) (2018): 516–524.e7. doi: 10.1016/j.cmet.2018.06.008.

30 M. Picard, J. Zhang, S. Hancock, et al. "Progressive Increase in mtDNA 3243A>G Heteroplasmy Causes Abrupt Transcriptional Reprogramming." *Proc Natl Acad Sci USA* 111(38) (2014): E4033–E4042. doi: 10.1073/pnas.1414028111.

31 A. Kasahara and L. Scorrano. "Mitochondria: From Cell Death Executioners to Regulators of Cell Differentiation." *Trends Cell Biol* 24(12) (2014): 761–770. doi: 10.1016/j.tcb.2014.08.005.

32 A. Kasahara, S. Cipolat, Y. Chen, G. W. Dorn, and L. Scorrano. "Mitochondrial Fusion Directs Cardiomyocyte Differentiation via Calcineurin and Notch Signaling." *Science* 342(6159) (2013): 734–737. doi: 10.1126/science.1241359.

33 Nikolaos Charmpilas and Nektarios Tavernarakis. "Mitochondrial Maturation Drives Germline Stem Cell Differentiation in *Caenorhabditis elegans*." *Cell Death Differ* 27(2) (2019). doi: 10.1038/s41418-019-0375-9.

34 Ryohei Iwata and Pierre Vanderhaeghen. "Regulatory Roles of Mitochondria and Metabolism in Neurogenesis." *Curr Opin Neurobiol* 69 (2021): 231–240. doi: 10.1016/j.conb.2021.05.003.

35 A. S. Rambold and J. Lippincott-Schwartz. "Mechanisms of Mitochondria and Autophagy Crosstalk." *Cell Cycle* 10(23) (2011): 4032–4038. doi: 10.4161/cc.10.23.18384.

36 Lane. *Power, Sex, Suicide.*

37 Jerry Edward Chipuk, Jarvier N. Mohammed, Jesse D. Gelles, and Yiyang Chen. "Mechanistic Connections Between Mitochondrial Biology and Regulated Cell Death." *Dev Cell* 56(9) (2021). doi: 10.1016/j.devcel.2021.03.033.

38 Lane. *Power, Sex, Suicide.*

Chapter 8

1 O. Lingjaerde. "Lactate-Induced Panic Attacks: Possible Involvement of Serotonin Reuptake Stimulation." *Acta Psychiatr Scand* 72(2) (985): 206–208. doi: 10.1111/j.1600-0447.1985.tb02596.x. PMID: 4050513.

2 M. B. First, W. C. Drevets, C. Carter, et al. "Clinical Applications of Neuroimaging in Psychiatric Disorders." *Am J Psychiatry* 175(9) (2018): 915–916. doi: 10.1176/appi.ajp.2018.1750701.

3 D. C. Wallace. "A Mitochondrial Etiology of Neuropsychiatric Disorders." *JAMA Psychiatry* 74(9) (2017): 863–864. doi: 10.1001/jamapsychiatry.2017.0397.

4 T. Kozicz, A. Schene, and E. Morava. "Mitochondrial Etiology of Psychiatric Disorders: Is This the Full Story?" *JAMA Psychiatry* 75(5) (2018): 527. doi: 10.1001/jamapsychiatry.2018.0018.

5 M. D. Brand and D. G. Nicholls. "Assessing Mitochondrial Dysfunction in Cells [published correction appears in *Biochem J* 437(3) (August 1, 2011): 575]. *Biochem J* 435(2) (2011): 297–312. doi: 10.1042/BJ20110162.

6 I. R. Lanza and K. S. Nair. "Mitochondrial Metabolic Function Assessed In Vivo and In Vitro." *Curr Opin Clin Nutr Metab Care* 13(5) (2010): 511–517. doi: 10.1097/MCO.0b013e32833cc93d.

7 A. H. De Mello, A. B. Costa, J. D. G. Engel, and G. T. Rezin. "Mitochondrial Dysfunction in Obesity." *Life Sci* 192 (2018): 26–32. doi: 10.1016/j.lfs.2017.11.019.

8 P. H. Reddy and M. F. Beal. "Amyloid Beta, Mitochondrial Dysfunction and Synaptic Damage: Implications for Cognitive Decline in Aging and Alzheimer's Disease." *Trends Mol Med* 14(2) (2008): 45–53. doi: 10.1016/j.molmed.2007.12.002.

9 Estela Area-Gomez, Ad de Groof, Eduardo Bonilla, Jorge Montesinos, Kurenai Tanji, Istvan Boldogh, Liza Pon, and Eric A. Schon. "A Key Role for MAM in Mediating Mitochondrial Dysfunction in Alzheimer Disease." *Cell Death Dis* 9(3) (2018): 335. doi: 10.1038/s41419-017-0215-0; R. H. Swerdlow. "Mitochondria and Mitochondrial Cascades in Alzheimer's Disease." *J Alzheimers Dis* 62(3) (2018): 1403–1416. doi: 10.3233/JAD-170585.

10 Fei Du, Xiao-Hong Zhu, Yi Zhang, Michael Friedman, Nanyin Zhang, Kâmil Uğurbil, and Wei Chen. "Tightly Coupled Brain Activity and Cerebral ATP Metabolic Rate." *Proc Natl Acad Sci USA* 105(17) (2008): 6409. doi: 10.1073/pnas.0710766105.

11 K. Todkar, H. S. Ilamathi, M. Germain. "Mitochondria and Lysosomes: Discovering Bonds." *Front Cell Dev Biol* 5 (2017):106. doi: 10.3389/fcell.2017.00106.

12 Q. Chu, T. F. Martinez, S. W. Novak, et al. "Regulation of the ER Stress Response by a MITOCHONDRIAL MICROPROTEIN." *Nat Commun* 10 (2019): 4883. doi: 10.1038/s41467-019-12816-z.

13 B. Kalman, F. D. Lublin, and H. Alder. "Impairment of Central and Peripheral Myelin in Mitochondrial Diseases." *Mult Scler* 2(6) (1997): 267–278. doi: 10.1177/135245859700200602; E. M. R. Lake, E. A. Steffler, C. D. Rowley, et al. "Altered Intracortical Myelin Staining in the Dorsolateral Prefrontal Cortex in Severe Mental Illness." *Eur Arch Psychiatry Clin Neurosci* 267 (2017): 369–376. doi: 10.1007/s00406-016-0730-5; J. Rice and C. Gu. "Function and Mechanism of Myelin Regulation in Alcohol Abuse and Alcoholism." *Bioessays* 41(7) (2019): e1800255. doi: 10.1002/bies.201800255. Epub May 16, 2019; Gerhard S. Drenthen, Walter H. Backes, Albert P. Aldenkamp, R. Jeroen Vermeulen, Sylvia Klinkenberg, and Jacobus F. A. Jansen. "On the Merits of Non-Invasive Myelin Imaging in Epilepsy, a Literature Review." *J Neurosci Methods* 338 (2020): 108687. doi: 10.1016/j.jneumeth.2020.108687; E. Papuć and K. Rejdak. "The Role of Myelin Damage in Alzheimer's Disease Pathology." *Arch Med Sci* 16(2) (2018): 345–351. doi: 10.5114/

aoms.2018.76863; G. Cermenati, F. Abbiati, S. Cermenati, et al. "Diabetes-Induced Myelin Abnormalities Are Associated with an Altered Lipid Pattern: Protective Effects of LXR Activation." *J Lipid Res* 53(2) (2012): 300–310. doi: 10.1194/jlr.M021188; M. Bouhrara, N. Khattar, P. Elango, et al. "Evidence of Association Between Obesity and Lower Cerebral Myelin Content in Cognitively Unimpaired Adults." *Int J Obes* (Lond) 45(4) (2021): 850–859. doi: 10.1038/s41366-021-00749-x.

14 A. Ebneth, R. Godemann, K. Stamer, S. Illenberger, B. Trinczek, and E. Mandelkow. "Overexpression of Tau Protein Inhibits Kinesin-Dependent Trafficking of Vesicles, Mitochondria, and Endoplasmic Reticulum: Implications for Alzheimer's Disease." *J Cell Biol* 143(3) (1998): 777–794. doi: 10.1083/jcb.143.3.777.

15 A. Cheng, J. Wang, N. Ghena, Q. Zhao, I. Perone, T. M. King, R. L. Veech, M. Gorospe, R. Wan, and M. P. Mattson. "SIRT3 Haploinsufficiency Aggravates Loss of GABAergic Interneurons and Neuronal Network Hyperexcitability in an Alzheimer's Disease Model." *J Neurosci* 40(3) (2020): 694–709. doi: 10.1523/JNEUROSCI.1446-19.2019.

16 J. Mertens, et al. "Differential Responses to Lithium in Hyperexcitable Neurons from Patients with Bipolar Disorder." *Nature* 527(7576) (2015): 95–99. doi: 10.1038/nature15526.

17 J. A. Rosenkranz, E. R. Venheim, and M. Padival. "Chronic Stress Causes Amygdala Hyperexcitability in Rodents." *Biol Psychiatry* 67(12) (2010): 1128–1136. doi: 10.1016/j.biopsych.2010.02.008.

18 Marco Morsch, Rowan Radford, Albert Lee, Emily Don, Andrew Badrock, Thomas Hall, Nicholas Cole, and Roger Chung. "In Vivo Characterization of Microglial Engulfment of Dying Neurons in the Zebrafish Spinal Cord." *Front Cell Neurosci* 9 (2015): 321. doi: 10.3389/fncel.2015.00321.

19 D. Alnæs, T. Kaufmann, D. van der Meer, et al. "Brain Heterogeneity in Schizophrenia and Its Association with Polygenic Risk." *JAMA Psychiatry* 76(7) (published online April 10, 2019): 739–748. doi: 10.1001/jamapsychiatry.2019.0257.

20 J. Allen, R. Romay-Tallon, K. J. Brymer, H. J. Caruncho, and L. E. Kalynchuk. "Mitochondria and Mood: Mitochondrial Dysfunction as a Key Player in the Manifestation of Depression." *Front Neurosci* 12 (June 6, 2018): 386. doi: 10.3389/fnins.2018.00386.

21 D. Ben-Shachar and R. Karry. "Neuroanatomical Pattern of Mitochondrial Complex I Pathology Varies Between Schizophrenia, Bipolar Disorder and Major Depression." *PLoS One* 3(11) (2008): e3676. doi: 10.1371/journal.pone.0003676.

22 J. Pu, Y. Liu, H. Zhang, et al. "An Integrated Meta-Analysis of Peripheral Blood Metabolites and Biological Functions in Major Depressive Disorder." *Mol Psychiatry* 26 (2020): 4265–4276. doi: 10.1038/s41380-020-0645-4.

23 C. Nasca, B. Bigio, F. S. Lee, et al. "Acetyl-L-Carnitine Deficiency in Patients with Major Depressive Disorder." *Proc Natl Acad Sci USA* 115(34) (2018): 8627–8632. doi: 10.1073/pnas.1801609115.

24 Ait Tayeb, Abd El Kader, Romain Colle, Khalil El-Asmar, Kenneth Chappell, Cécile Acquaviva-Bourdain, Denis J. David, Séverine Trabado, et al. "Plasma Acetyl-L-Carnitine and L-Carnitine in Major Depressive Episodes: A Case–Control Study Before and After Treatment." *Psychol Med* (2021): 1–10. doi: 10.1017/S003329172100413X.

25 E. Gebara, O. Zanoletti, S. Ghosal, J. Grosse, B. L. Schneider, G. Knott, S. Astori, and C. Sandi. "Mitofusin-2 in the Nucleus Accumbens Regulates Anxiety and Depression-like Behaviors Through Mitochondrial and Neuronal Actions." *Biol Psychiatry* 89(11) (2021): 1033–1044. doi: 10.1016/j.biopsych.2020.12.003.

26 M. D. Altschule, D. H. Henneman, P. Holliday, and R. M. Goncz. "Carbohydrate Metabolism in Brain Disease. VI. Lactate Metabolism After Infusion of Sodium D-Lactate in Manic-Depressive and Schizophrenic Psychoses." *AMA Arch Intern Med* 98 (1956): 35–38. doi: 10.1001/archinte.1956.00250250041006.

27 Gerwyn Morris, Ken Walder, Sean L. McGee, Olivia M. Dean, Susannah J. Tye, Michael Maes, and Michael Berk. "A Model of the Mitochondrial Basis of Bipolar Disorder." *Neurosci Biobehav Rev* 74 (2017): 1–20. doi: 10.1016/j.neubiorev.2017.01.014.

28 Anna Giménez-Palomo, Seetal Dodd, Gerard Anmella, Andre F. Carvalho, Giselli Scaini, Joao Quevedo, Isabella Pacchiarotti, Eduard Vieta, and Michael Berk. "The Role of Mitochondria in Mood Disorders: From Physiology to Pathophysiology and to Treatment." *Front Psychiatry* 12 (2021): 977. doi: 10.3389/fpsyt.2021.546801.

29 D. Wang, Z. Li, W. Liu, et al. "Differential Mitochondrial DNA Copy Number in Three Mood States of Bipolar Disorder." *BMC Psychiatry* 18 (2018): 149. doi: 10.1186/s12888-018-1717-8.

30 G. Preston, F. Kirdar, and T. Kozicz. "The Role of Suboptimal Mitochondrial Function in Vulnerability to Post-traumatic Stress Disorder." *J Inherit Metab Dis* 41(4) (2018): 585–596. doi: 10.1007/s10545-018-0168-1.

31 S. Ali, M. Patel, S. Jabeen, R. K. Bailey, T. Patel, M. Shahid, W. J. Riley, and A. Arain. "Insight into Delirium." *Innov Clin Neurosci* 8(10) (2011): 25–34. PMID: 22132368.

32 A. J. Slooter, D. Van, R. R. Leur, and I. J. Zaal. "Delirium in Critically Ill Patients." *Handb Clin Neurol* 141 (2017): 449–466. doi: 10.1016/B978-0-444-63599-0.00025-9.

33 G. L. Engel and J. Romano. "Delirium, a Syndrome of Cerebral Insufficiency." *J Chronic Dis.* 9(3) (1959): 260–277. doi: 10.1016/0021-9681(59)90165-1.

34 J. E. Wilson, M. F. Mart, C. Cunningham, et al. "Delirium." *Nat Rev Dis Primers* 6 (2020): 90. doi: 10.1038/s41572-020-00223-4.

35 L. R. Haggstrom, J. A. Nelson, E. A. Wegner, and G. A. Caplan. "2-(18)F-fluoro-2-deoxyglucose Positron Emission Tomography in Delirium." *J. Cereb. Blood Flow Metab* 37(11) (2017): 3556–3567. doi: 10.1177/0271678X17701764.

36 A. J. Slooter, D. Van, R. R. Leur, and I. J. Zaal. "Delirium in Critically Ill Patients." *Handb Clin Neurol* 141 (2017): 449–466. doi: 10.1016/B978-0-444-63599-0.00025-9; T. E. Goldberg, C. Chen, Y. Wang, et al. "Association of Delirium with Long-Term Cognitive Decline: A Meta-analysis." *JAMA Neurol.* Published online July 13, 2020. doi: 10.1001/jamaneurol.2020.2273.

37 G. Naeije, I. Bachir, N. Gaspard, B. Legros, and T. Pepersack. "Epileptic Activities and Older People Delirium." *Geriatr Gerontol Int* 14(2) (2014): 447–451. doi: 10.1111/ggi.12128.

38 Jorge I. F. Salluh, Han Wang, Eric B. Schneider, Neeraja Nagaraja, Gayane Yenokyan, Abdulla Damluji, Rodrigo B. Serafim, and Robert D. Stevens. "Outcome of Delirium in Critically Ill Patients: Systematic Review and Meta-Analysis." *BMJ* 350 (2015). doi: 10.1136/bmj.h2538.

39 Sharon K. Inouye. "Delirium in Older Persons." *N Engl J Med* 354(11) (2006): 1157–65. doi: 10.1056/NEJMra052321.

40 Robert Hatch, Duncan Young, Vicki Barber, John Griffiths, David A. Harrison, and Peter Watkinson. "Anxiety, Depression and Post Traumatic Stress Disorder after Critical Illness: A UK-Wide Prospective Cohort Study." *Crit Care* 22(1) (2018): 310. doi: 10.1186/s13054-018-2223-6.

41 O. Plana-Ripoll, C. B. Pedersen, Y. Holtz, et al. "Exploring Comorbidity Within Mental Disorders Among a Danish National Population." *JAMA Psychiatry* (published online January 16, 2019). doi: 10.1001/jamapsychiatry.2018.3658ArticleGoogle Scholar.

Chapter 10

1 S. Umesh and S. H. Nizamie. "Genetics in Psychiatry." *Indian J Hum Genet* 20(2) (2014): 120–128. doi: 10.4103/0971-6866.142845.

2 Richard Border, Emma C. Johnson, Luke M. Evans, Andrew Smolen, Noah Berley, Patrick F. Sullivan, and Matthew C. Keller. "No Support for Historical Candidate Gene or Candidate Gene-by-Interaction Hypotheses for Major Depression Across Multiple Large Samples." *Am J Psychiatry* 176(5) (2019): 376–387. doi: 10.1176/appi.ajp.2018.18070881.

3 G. Scaini, G. T. Rezin, A. F. Carvalho, E. L. Streck, M. Berk, and J. Quevedo. "Mitochondrial Dysfunction in Bipolar Disorder: Evidence, Pathophysiology and Translational Implications." *Neurosci Biobehav* 68 (Rev. September 2016): 694–713. doi: 10.1016/j.neubiorev.2016.06.040.

4 S. Michels, G. K. Ganjam, H. Martins, et al. "Downregulation of the Psychiatric Susceptibility Gene *Cacna1c* Promotes Mitochondrial Resilience to Oxidative Stress in Neuronal Cells." *Cell Death Dis* 4(54) (2018): 54. doi: 10.1038/s41420-018-0061-6.

5 Lixia Qin, Zhu Xiongwei, and Robert P. Friedland. "ApoE and Mitochondrial Dysfunction." *Neurology* 94(23) (2020): 1009. doi: 10.1212/WNL.0000000000009569.

6 Y. Yamazaki, N. Zhao, T. R. Caulfield, C. C. Liu, and G. Bu. "Apolipoprotein E and Alzheimer Disease: Pathobiology and Targeting Strategies." *Nat Rev Neurol* 15(9) (2019): 501–518. doi: 10.1038/s41582-019-0228-7.

7 J. Yin, E. M. Reiman, T. G. Beach, et al. "Effect of ApoE Isoforms on Mitochondria in Alzheimer Disease." *Neurology* 94(23) (2020): e2404–e2411. doi: 10.1212/WNL.0000000000009582.

8 E. Schmukler, S. Solomon, S. Simonovitch, et al. "Altered Mitochondrial Dynamics and Function in *APOE4*-Expressing Astrocytes." *Cell Death Dis* 11(7) (2020): 578. doi: 10.1038/s41419-020-02776-4.

9 A. L. Lumsden, A. Mulugeta, A. Zhou, and E. Hyppönen. "Apolipoprotein E (APOE) Genotype-Associated Disease Risks: A Phenome-Wide, Registry-Based, Case-Control Study Utilising the UK Biobank." *EBioMedicine* 59 (2020):102954. doi: 10.1016/j.ebiom.2020.102954.

10 M. S. Sharpley, C. Marciniak, K. Eckel-Mahan, M. McManus, M. Crimi, K. Waymire, C. S. Lin, S. Masubuchi, N. Friend, M. Koike, D. Chalkia, G. MacGregor, P. Sassone-Corsi, and D. C. Wallace. "Heteroplasmy of Mouse mtDNA Is Genetically Unstable and Results in Altered Behavior and Cognition." *Cell* 151(2) (2012): 333–343. doi: 10.1016/j.cell.2012.09.004. PMID: 23063123; PMCID: PMC4175720.

11 Centers for Disease Control and Prevention. "What Is Epigenetics?" CDC, US Department of Health and Human Services. https://www.cdc.gov/genomics/disease/epigenetics.htm. Retrieved 10/30/21.

12 T. J. Roseboom. "Epidemiological Evidence for the Developmental Origins of Health and Disease: Effects of Prenatal Undernutrition in Humans." *J Endocrinol* 242(1) (July 1, 2019): T135–T144. doi: 10.1530/JOE-18-0683.

13 J. P. Etchegaray and R. Mostoslavsky. "Interplay Between Metabolism and Epigenetics: A Nuclear Adaptation to Environmental Changes." *Mol Cell* 62(5) (2016): 695–711. doi: 10.1016/j.molcel.2016.05.029.

14 P. H. Ear, A. Chadda, S. B. Gumusoglu, M. S. Schmidt, S. Vogeler, J. Malicoat, J. Kadel, M. M. Moore, M. E. Migaud, H. E. Stevens, and C. Brenner. "Maternal Nicotinamide Riboside Enhances Postpartum Weight Loss, Juvenile Offspring Development, and Neurogenesis of Adult Offspring." *Cell Rep* 26(4) (2019): 969–983.e4. doi: 10.1016/j.celrep.2019.01.007.

15 R. Yehuda and A. Lehrner. "Intergenerational Transmission of Trauma Effects: Putative Role of Epigenetic Mechanisms." *World Psychiatry* 17(3) (2018): 243–257. doi: 10.1002/wps.20568.

16 D. A. Dickson, J. K. Paulus, V. Mensah, et al. "Reduced Levels of miRNAs 449 and 34 in Sperm of Mice and Men Exposed to Early Life Stress." *Transl Psychiatry* 8 (2018): 101. doi: 10.1038/s41398-018-0146-2.

17 S. Lupien, B. McEwen, M. Gunnar, et al. "Effects of Stress Throughout the Lifespan on the Brain, Behaviour, and Cognition." *Nat Rev Neurosci* 10 (2009): 434–445. doi: 10.1038/nrn2639.

Chapter 11

1 Julian M. Yabut, Justin D. Crane, Alexander E. Green, Damien J. Keating, Waliul I. Khan, and Gregory R. Steinberg. "Emerging Roles for Serotonin in Regulating Metabolism: New Implications for an Ancient Molecule." *Endocr Rev* 40(4) (2019): 1092–1107. doi: 10.1210/er.2018-00283.

2 Sashaina E. Fanibunda, Deb Sukrita, Babukrishna Maniyadath, Praachi Tiwari, Utkarsha Ghai, Samir Gupta, Dwight Figueiredo, et al. "Serotonin Regulates Mitochondrial Biogenesis and Function in Rodent Cortical Neurons via the 5-HT_{2A} Receptor and SIRT1–PGC-1α Axis." *Proc Natl Acad Sci USA* 116(22) (2019): 11028. doi: 10.1073/pnas.1821332116.

3 M. Accardi, B. Daniels, P. Brown, et al. "Mitochondrial Reactive Oxygen Species Regulate the Strength of Inhibitory GABA-Mediated Synaptic Transmission." *Nat Commun* 5 (2014): 3168. doi: 10.1038/ncomms4168.

4 A. K. Kanellopoulos, V. Mariano, M. Spinazzi, Y. J. Woo, C. McLean, U. Pech, K. W. Li, J. D. Armstrong, A. Giangrande, P. Callaerts, A. B. Smit, B. S. Abrahams, A. Fiala, T. Achsel, and C. Bagni. "Aralar Sequesters GABA into Hyperactive Mitochondria, Causing Social Behavior Deficits." *Cell* 180(6) (March 19, 2020): 1178–1197.e20. doi: 10.1016/j.cell.2020.02.044.

5 Ryutaro Ikegami, Ippei Shimizu, Takeshi Sato, Yohko Yoshida, Yuka Hayashi, Masayoshi Suda, Goro Katsuumi, et al. "Gamma-Aminobutyric Acid Signaling in Brown Adipose Tissue Promotes Systemic Metabolic Derangement in Obesity." *Cell Rep* 24(11) (2018): 2827–2837.e5. doi: 10.1016/j.celrep.2018.08.024.

6 S. M. Graves, Z. Xie, K. A. Stout, et al. "Dopamine Metabolism by a Monoamine Oxidase Mitochondrial Shuttle Activates the Electron Transport Chain." *Nat Neurosci* 23 (2020): 15–20.

7 D. Aslanoglou, S. Bertera, M. Sánchez-Soto, et al. "Dopamine Regulates Pancreatic Glucagon and Insulin Secretion via Adrenergic and Dopaminergic Receptors." *Transl Psychiatry* 11(1) (2021): 59. doi: 10.1038/s41398-020-01171-z.

8 M. van der Kooij, F. Hollis, L. Lozano, et al. "Diazepam Actions in the VTA Enhance Social Dominance and Mitochondrial Function in the Nucleus Accumbens by Activation of Dopamine D1 Receptors." *Mol Psychiatry* 23(3) (2018): 569–578. doi: 10.1038/mp.2017.135.

9 M. van der Kooij, et al. "Diazepam Actions in the VTA Enhance Social Dominance and Mitochondrial Function in the Nucleus Accumbens by Activation of Dopamine D1 Receptors."

10 T. L. Emmerzaal, G. Nijkamp, M. Veldic, S. Rahman, A. C. Andreazza, E. Morava, R. J. Rodenburg, and T. Kozicz. "Effect of Neuropsychiatric Medications on Mitochondrial Function: For Better or for Worse." *Neurosci Biobehav* 127 (Rev. August 2021): 555–571. doi: 10.1016/j.neubiorev.2021.05.001.

11 Martin Lundberg, Vincent Millischer, Lena Backlund, Lina Martinsson, Peter Stenvinkel, Carl M. Sellgren, Catharina Lavebratt, and Martin Schalling. "Lithium and the Interplay Between Telomeres and Mitochondria in Bipolar Disorder." *Front Psychiatry* 11 (2020): 997. doi: 10.3389/fpsyt.2020.586083.

12 M. Hu, R. Wang, X. Chen, M. Zheng, P. Zheng, Z. Boz, R. Tang, K. Zheng, Y. Yu, and X. F. Huang. "Resveratrol Prevents Haloperidol-Induced Mitochondria Dysfunction Through the Induction of Autophagy in SH-SY5Y Cells." *Neurotoxicology* 87 (2021): 231–242. doi: 10.1016/j.neuro.2021.10.007.

13 D. C. Goff, G. Tsai, M. F. Beal, and J. T. Coyle. "Tardive Dyskinesia and Substrates of Energy Metabolism in CSF." *Am J Psychiatry* 152(12) (1995): 1730–6. doi: 10.1176/ajp.152.12.1730. PMID: 8526238.

14 M. Salsaa, B.Pereira, J. Liu, et al. "Valproate Inhibits Mitochondrial Bioenergetics and Increases Glycolysis in *Saccharomyces cerevisiae*." *Sci Rep* 10(1) (2020): 11785. doi: 10.1038/s41598-020-68725-5.

15 J. F. Hayes, A. Lundin, S. Wicks, G. Lewis, I. C. K. Wong, D. P. J. Osborn, and C. Dalman. "Association of Hydroxylmethyl Glutaryl Coenzyme A Reductase Inhibitors, L-Type Calcium Channel Antagonists, and Biguanides with Rates of Psychiatric Hospitalization and Self-Harm in Individuals with Serious Mental Illness." *JAMA Psychiatry* 76(4) (2019): 382–390. doi: 10.1001/jamapsychiatry.2018.3907.

16 S. Martín-Rodríguez, P. de Pablos-Velasco, and J. A. L. Calbet. "Mitochondrial Complex I Inhibition by Metformin: Drug-Exercise Interactions." *Trends Endocrinol Metab* 31(4) (April 2020): 269–271. doi: 10.1016/j.tem.2020.02.003.

Chapter 12

1 P. Maechler. "Mitochondrial Function and Insulin Secretion." *Mol Cell Endocrinol* 379(1–2) (2013): 12–18. doi: 10.1016/j.mce.2013.06.019.

2 W. I. Sivitz and M. A. Yorek. "Mitochondrial Dysfunction in Diabetes: From Molecular Mechanisms to Functional Significance and Therapeutic Opportunities." *Antioxid Redox Signal* 12(4) (2010): 537–577. doi: 10.1089/ars.2009.2531.

3 C. S. Stump, K. R. Short, M. L. Bigelow, J. M. Schimke, and K. S. Nair. "Effect of Insulin on Human Skeletal Muscle Mitochondrial ATP Production, Protein Synthesis, and mRNA Transcripts." *Proc Natl Acad Sci USA* 100(13) (2003): 7996–8001. doi: 10.1073/pnas.1332551100.

4 A. Kleinridders, H. A. Ferris, W. Cai, and C. R. Kahn. "Insulin Action in Brain Regulates Systemic Metabolism and Brain Function." *Diabetes* 63(7) (2014): 2232–2243. doi: 10.2337/db14-0568.

5 E. Blázquez, E. Velázquez, V. Hurtado-Carneiro, and J. M. Ruiz-Albusac. "Insulin in the Brain: Its Pathophysiological Implications for States Related with Central Insulin Resistance, Type 2 Diabetes and Alzheimer's Disease." *Front Endocrinol* (Lausanne) 5 (2014): 161. doi: 10.3389/fendo.2014.00161.

6 Z. Jin, Y. Jin, S. Kumar-Mendu, E. Degerman, L. Groop, and B. Birnir. "Insulin Reduces Neuronal Excitability by Turning on GABA(A) Channels That Generate Tonic Current." *PLoS One* 6(1) (2011): e16188. doi: 10.1371/journal.pone.0016188.

7 Ismael González-García, Tim Gruber, and Cristina García-Cáceres. "Insulin Action on Astrocytes: From Energy Homeostasis to Behaviour." *J Neuroendocrinol* 33(4) (2021): e12953. doi: 10.1111/jne.12953.

8 A. Kleinridders, W. Cai, L. Cappellucci, A. Ghazarian, W. R. Collins, S. G. Vienberg, E. N. Pothos, and C. R. Kahn. "Insulin Resistance in Brain Alters Dopamine Turnover and Causes Behavioral Disorders." *Proc Natl Acad Sci USA* 112(11) (2015): 3463–3468. doi: 10.1073/pnas.1500877112.

9 Virginie-Anne Chouinard, David C. Henderson, Chiara Dalla Man, Linda Valeri, Brianna E. Gray, Kyle P. Ryan, Aaron M. Cypess, Claudio Cobelli, Bruce M. Cohen, and Dost Öngür. "Impaired Insulin Signaling in Unaffected Siblings and Patients with First-Episode Psychosis." *Mol Psychiatry* 24 (2018). doi: 10.1038/s41380-018-0045-1.

10 B. I. Perry, J. Stochl, R. Upthegrove, et al. "Longitudinal Trends in Childhood Insulin Levels and Body Mass Index and Associations with Risks of Psychosis and Depression in Young Adults." *JAMA Psychiatry* 78(4) (2021): 416–425. doi: 10.1001/jamapsychiatry.2020.4180.

11 B. J. Neth and S. Craft. "Insulin Resistance and Alzheimer's Disease: Bioenergetic Linkages. *Front Aging Neurosci* 9 (2017): 345. doi: 10.3389/fnagi.2017.00345; Y. An, V. R. Varma, S. Varma, R. Casanova, E. Dammer, O. Pletnikova, C. W. Chia, J. M. Egan, L. Ferrucci, J. Troncoso, A. I. Levey, J. Lah, N. T. Seyfried, C. Legido-Quigley, R. O'Brien, and M. Thambisetty. "Evidence for Brain Glucose Dysregulation in Alzheimer's Disease." *Alzheimers Dement* 14(3) (2018): 318–329. doi: 10.1016/j.jalz.2017.09.011.

12 S. Craft, L. D. Baker, T. J. Montine, et al. "Intranasal Insulin Therapy for Alzheimer Disease and Amnestic Mild Cognitive Impairment: A Pilot Clinical Trial." *Arch Neurol* 69(1) (2012): 29–38. doi: 10.1001/archneurol.2011.233.

13 S. Craft, R. Raman, T. W. Chow, et al. "Safety, Efficacy, and Feasibility of Intranasal Insulin for the Treatment of Mild Cognitive Impairment and Alzheimer Disease Dementia: A Randomized Clinical Trial." *JAMA Neurol* 77(9) (2020): 1099–1109. doi: 10.1001/jamaneurol.2020.1840.

14 R. S. McIntyre, J. K. Soczynska, H. O. Woldeyohannes, A. Miranda, A. Vaccarino, G. Macqueen, G. F. Lewis, and S. H. Kennedy. "A Randomized, Double-Blind, Controlled Trial Evaluating the Effect of Intranasal Insulin on Neurocognitive Function in Euthymic Patients with Bipolar Disorder." *Bipolar Disord* 14(7) (2012): 697–706. doi: 10.1111/bdi.12006.

15 Jamaica R. Rettberg, Jia Yao, and Roberta Diaz Brinton. "Estrogen: A Master Regulator of Bioenergetic Systems in the Brain and Body." *Front Neuroendocrinol* 35(1) (2014): 8–30. doi: 10.1016/j.yfrne.2013.08.001.

16 L. Mosconi, V. Berti, C. Quinn, P. McHugh, G. Petrongolo, R. S. Osorio, C. Connaughty, A. Pupi, S. Vallabhajosula, R. S. Isaacson, M. J. de Leon, R. H. Swerdlow, and R. D. Brinton. "Perimenopause and Emergence of an Alzheimer's Bioenergetic Phenotype in Brain and Periphery." *PLoS One* 12(10) (2017): e0185926. doi: 10.1371/journal.pone.0185926

17 Y. Hara, F. Yuk, R. Puri, W. G. Janssen, P. R. Rapp, and J. H. Morrison. "Presynaptic Mitochondrial Morphology in Monkey Prefrontal Cortex Correlates with Working Memory and Is Improved with Estrogen Treatment." *Proc Natl Acad Sci USA* 111(1) (2014): 486–491. doi: 10.1073/pnas.1311310110.

18 Charlotte Wessel Skovlund, Lina Steinrud Mørch, Lars Vedel Kessing, and Øjvind Lidegaard. "Association of Hormonal Contraception with Depression." *JAMA Psychiatry* 73(11) (2016): 1154–1162. doi: 10.1001/jamapsychiatry.2016.2387.

19 C. W. Skovlund, L. S. Mørch, L. V. Kessing, T. Lange, and Ø. Lidegaard. "Association of Hormonal Contraception with Suicide Attempts and Suicides." *Am J Psychiatry* 175(4) (2018): 336–342. doi: 10.1176/appi.ajp.2017.17060616.

20 Federica Cioffi, Rosalba Senese, Antonia Lanni, and Fernando Goglia. "Thyroid Hormones and Mitochondria: With a Brief Look at Derivatives and Analogues." *Mol Cell Endocrinol* 379(1) (2013): 51–61. doi: 10.1016/j.mce.2013.06.006.

21 Rohit A. Sinha, Brijesh K. Singh, Jin Zhou, Yajun Wu, Benjamin L. Farah, Kenji Ohba, Ronny Lesmana, Jessica Gooding, Boon-Huat Bay, and Paul M. Yen. "Thyroid Hormone Induction of Mitochondrial Activity Is Coupled to Mitophagy via ROS-AMPK-ULK1 Signaling." *Autophagy* 11(8) (2015): 1341–1357. doi: 10.1080/15548627.2015.1061849.

22 S. Chakrabarti. "Thyroid Functions and Bipolar Affective Disorder." *J Thyroid Res* 2011 (2011): 306367. doi: 10.4061/2011/306367; N. C. Santos, P. Costa, D. Ruano, et al. "Revisiting Thyroid Hormones in Schizophrenia." *J Thyroid Res* 2012 (2012): 569147. doi: 10.1155/2012/569147.

Chapter 13

1 Steven W. Cole, John P. Capitanio, Katie Chun, Jesusa M. G. Arevalo, Jeffrey Ma, and John T. Cacioppo. "Myeloid Differentiation Architecture of Leukocyte Transcriptome Dynamics in Perceived Social Isolation." *Proc Natl Acad Sci USA* 112(49) (2015): 15142–15147. doi: 10.1073/pnas.1514249112.

2 Y. Luo, L. C. Hawkley, L. J. Waite, and J. T. Cacioppo. "Loneliness, Health, and Mortality in Old Age: A National Longitudinal Study." *Soc Sci Med* 74(6) (2012): 907–914. doi: 10.1016/j.socscimed.2011.11.028.

3 J. Wang, D. Xiao, H. Chen, et al. "Cumulative Evidence for Association of Rhinitis and Depression." *Allergy Asthma Clin Immunol* 17(1) (2021): 111. doi: 10.1186/s13223-021-00615-5.

4 O. Köhler-Forsberg, L. Petersen, C. Gasse, et al. "A Nationwide Study in Denmark of the Association Between Treated Infections and the Subsequent Risk of Treated Mental Disorders in Children and Adolescents." *JAMA Psychiatry* 76(3) (2019): 271–279. doi: 10.1001/jamapsychiatry.2018.3428.

5 A. West, G. Shadel, and Ghosh. "Mitochondria in Innate Immune Responses." *Nat Rev Immunol* 11 (2011): 389–402. doi: 10.1038/nri2975

6 Z. Liu and T. S. Xiao. "Partners with a Killer: Metabolic Signaling Promotes Inflammatory Cell Death." *Cell* 184(17) (2021): 4374–4376. doi: 10.1016/j.cell.2021.07.036.

7 D. N. Doll, S. L. Rellick, T. L. Barr, X. Ren, and J. W. Simpkins. "Rapid Mitochondrial Dysfunction Mediates TNF-Alpha-Induced Neurotoxicity." *J Neurochem* 132(4) (2015): 443–451. doi: 10.1111/jnc.13008.

8 B. Shan, E. Vazquez, and J. A. Lewis. "Interferon Selectively Inhibits the Expression of Mitochondrial Genes: A Novel Pathway for Interferon-Mediated Responses." *EMBO J* 9(13) (1990): 4307–4314. doi: 10.1002/j.1460-2075.1990.tb07879.x.

9 S. B. Minchenberg and P. T. Massa. "The Control of Oligodendrocyte Bioenergetics by Interferon-Gamma (IFN-γ) and Src Homology Region 2 Domain-Containing Phosphatase-1 (SHP-1)." *J Neuroimmunol* 331 (2019): 46–57. doi: 10.1016/j.jneuroim.2017.10.015.

10 H. G. Coman, D. C. Herţa, and B. Nemeş. "Psychiatric Adverse Effects Of Interferon Therapy." *Clujul Med* 86(4) (2013): 318–320.

11 B. J. S. Al-Haddad, B. Jacobsson, S. Chabra, D. Modzelewska, E. M. Olson, R. Bernier, D. A. Enquobahrie, H. Hagberg, S. Östling, L. Rajagopal, K. M. Adams Waldorf, and V. Sengpiel. "Long-Term Risk of Neuropsychiatric Disease After Exposure to Infection In Utero." *JAMA Psychiatry* 76(6) (2019): 594–602. doi: 10.1001/jamapsychiatry.2019.0029. PMID: 30840048; PMCID: PMC6551852.

12 A. H. Miller and C. L. Raison. "Are Anti-inflammatory Therapies Viable Treatments for Psychiatric Disorders? Where the Rubber Meets the Road." *JAMA Psychiatry* 72(6) (2015): 527–528. doi:10.1001/jamapsychiatry.2015.22.

Chapter 14

1 Jaqueline B. Schuch, Julia P. Genro, Clarissa R. Bastos, Gabriele Ghisleni, and Luciana Tovo-Rodrigues. "The Role of CLOCK Gene in Psychiatric Disorders: Evidence from Human and Animal Research." *Am J Med Genet Part B* 177(2) (2018): 181–198. doi: 10.1002/ajmg.b.32599.

2 Karen Schmitt, Amandine Grimm, Robert Dallmann, Bjoern Oettinghaus, Lisa Michelle Restelli, Melissa Witzig, Naotada Ishihara, et al. "Circadian Control of DRP1 Activity Regulates Mitochondrial Dynamics and Bioenergetics." *Cell Metab* 27(3) (2018): 657–666.e5. doi: 10.1016/j.cmet.2018.01.011.

3 Ana C. Andreazza, Monica L. Andersen, Tathiana A. Alvarenga, Marcos R. de-Oliveira, Fernanda Armani, Francieli S. Ruiz, Larriany Giglio, José C. F. Moreira, Flávio Kapczinski, and Sergio Tufik. "Impairment of the Mitochondrial Electron Transport Chain Due to Sleep Deprivation in Mice." *J Psychiatr Res* 44(12) (2010): 775–780. doi: 10.1016/j.jpsychires.2010.01.015.

4 Martin Picard, Bruce S. McEwen, Elissa S. Epel, and Carmen Sandi. "An Energetic View of Stress: Focus on Mitochondria." *Front Neuroendocrinol* 49 (2018): 72–85. doi: 10.1016/j.yfrne.2018.01.001.

5 Chongyang Chen, Chao Yang, Jing Wang, Xi Huang, Haitao Yu, Shangming Li, Shupeng Li, et al. "Melatonin Ameliorates Cognitive Deficits Through Improving Mitophagy in a Mouse Model of Alzheimer's Disease." *J Pineal Res* 71(4) (2021): e12774. doi: 10.1111/jpi.12774.

6 H. Zhao, H. Wu, J. He, et al. "Frontal Cortical Mitochondrial Dysfunction and Mitochondria-Related β-Amyloid Accumulation by Chronic Sleep Restriction in Mice." *Neuroreport* 27(12) (2016): 916–922. doi: 10.1097/WNR.0000000000000631.

7 C. B. Peek, A. H. Affinati, K. M. Ramsey, H. Y. Kuo, W. Yu, L. A. Sena, O. Ilkayeva, B. Marcheva, Y. Kobayashi, C. Omura, D. C. Levine, D. J. Bacsik, D. Gius, C. B. Newgard, E. Goetzman, N. S. Chandel, J. M. Denu, M. Mrksich, and J. Bass. "Circadian Clock NAD+ Cycle Drives Mitochondrial Oxidative Metabolism in Mice." *Science* 342(6158) (2013): 1243417. doi: 10.1126/science.1243417.

8 A. Kempf, S. M. Song, C. B. Talbot, et al. "A Potassium Channel β-subunit Couples Mitochondrial Electron Transport to Sleep." *Nature* 568(7751) (2019): 230–234. doi: 10.1038/s41586-019-1034-5.

9 Keri J. Fogle, Catherina L. Mobini, Abygail S. Paseos, and Michael J. Palladino. "Sleep and Circadian Defects in a Drosophila Model of Mitochondrial Encephalomyopathy." *Neurobiol Sleep Circadian Rhythm* 6 (2019): 44–52. doi: 10.1016/j.nbscr.2019.01.003.

10 Guido Primiano, Valerio Brunetti, Catello Vollono, Anna Losurdo, Rossana Moroni, Giacomo Della Marca, and Serenella Servidei. "Sleep-Disordered Breathing in Adult Patients with Mitochondrial Diseases." *Neurology* 96(2) (2021): e241. doi: 10.1212/WNL.0000000000011005.

11 N. N. Osborne, C. Núñez-Álvarez, S. Del Olmo-Aguado, and J. Merrayo-Lloves. "Visual Light Effects on Mitochondria: The Potential Implications in Relation to Glaucoma." *Mitochondrion* 36 (2017): 29–35. doi: 10.1016/j.mito.2016.11.009. Epub 2016 Nov 24. PMID: 27890822.

12 A. Sreedhar, L. Aguilera-Aguirre, and K. K. Singh. "Mitochondria in Skin Health, Aging, and Disease." *Cell Death Dis* 11(6) (2020): 444. doi: 10.1038/s41419-020-2649-z.

13 H. Zhu, N. Wang, L. Yao, Q. Chen, R. Zhang, J. Qian, Y. Hou, W. Guo, S. Fan, S. Liu, Q. Zhao, F. Du, X. Zuo, Y. Guo, Y. Xu, J. Li, T. Xue, K. Zhong, X. Song, G. Huang, and W. Xiong. "Moderate UV Exposure Enhances Learning and Memory by Promoting a Novel Glutamate Biosynthetic Pathway in the Brain." *Cell* 173(7) (2018): 1716–1727.e17. doi: 10.1016/j.cell.2018.04.014.

14 F. Salehpour, J. Mahmoudi, F. Kamari, S. Sadigh-Eteghad, S. H. Rasta, and M. R. Hamblin. "Brain Photobiomodulation Therapy: A Narrative Review." *Mol Neurobiol* 55(8) (2018): 6601–6636. doi: 10.1007/s12035-017-0852-4.

15 P. D. Campbell, A. M. Miller, and M. E. Woesner. "Bright Light Therapy: Seasonal Affective Disorder and Beyond." *Einstein J Biol Med* 32 (2017): E13–E25. PMID: 31528147; PMCID: PMC6746555.

16 R. Noordam, et al. "Bright Sunlight Exposure May Decrease the Risk for Diabetes and CVD." *J Clin Endocrinol Metab* 104(7) (2019): 2903–2910. doi: 10.1210/jc.2018-02532.

17 J. F. Gottlieb, F. Benedetti, P. A. Geoffroy, T. E. G. Henriksen, R. W. Lam, G. Murray, J. Phelps, D. Sit, H. A. Swartz, M. Crowe, B. Etain, E. Frank, N. Goel, B. C. M. Haarman, M. Inder, H. Kallestad, S. Jae Kim, K. Martiny, Y. Meesters, R. Porter, R. F. Riemersma-van der Lek, P. S. Ritter, P. F. J. Schulte, J. Scott, J. C. Wu, X. Yu, and S. Chen. "The Chronotherapeutic Treatment of Bipolar Disorders: A Systematic Review and Practice Recommendations from the ISBD Task Force on Chronotherapy and Chronobiology." *Bipolar Disord* 21(8) (2019): 741–773. doi: 10.1111/bdi.12847.

Chapter 15

1 N. D. Volkow, R. A. Wise, and R. Baler. "The Dopamine Motive System: Implications for Drug and Food Addiction." *Nat Rev Neurosci* 18(12) (2017): 741–752. doi: 10.1038/nrn.2017.130.

2 W. Li, Z. Wang, S. Syed, et al. "Chronic Social Isolation Signals Starvation and Reduces Sleep in Drosophila." *Nature* 597(7875) (2021): 239–244. doi: 10.1038/s41586-021-03837-0.

3 G. Xia, Y. Han, F. Meng, et al. "Reciprocal Control of Obesity and Anxiety–Depressive Disorder via a GABA and Serotonin Neural Circuit." *Mol Psychiatry* 26(7) (2021): 2837–2853. doi: 10.1038/s41380-021-01053-w.

4 E. Ginter and V. Simko. "New Data on Harmful Effects of Trans-Fatty Acids." *Bratisl Lek Listy* 117(5) (2016): 251–253. doi: 10.4149/bll_2016_048.

5 C. S. Pase, V. G. Metz, K. Roversi, K. Roversi, L. T. Vey, V. T. Dias, C. F. Schons, C. T. de David Antoniazzi, T. Duarte, M. Duarte, and M. E. Burger. "Trans Fat Intake During Pregnancy or Lactation Increases Anxiety-like

Behavior and Alters Proinflammatory Cytokines and Glucocorticoid Receptor Levels in the Hippocampus of Adult Offspring." *Brain Res Bull* 166 (2021): 110–117. doi: 10.1016/j.brainresbull.2020.11.016.

6 Theodora Psaltopoulou, Theodoros N. Sergentanis, Demosthenes B. Panagiotakos, Ioannis N. Sergentanis, Rena Kosti, and Nikolaos Scarmeas. "Mediterranean Diet, Stroke, Cognitive Impairment, and Depression: A Meta-analysis." *Ann Neurol* 74(4) (2013): 580–91. doi: 10.1002/ana.23944.

7 M. P. Mollica, G. Mattace Raso, G. Cavaliere, et al. "Butyrate Regulates Liver Mitochondrial Function, Efficiency, and Dynamics in Insulin-Resistant Obese Mice." *Diabetes* 66(5) (2017): 1405–1418. doi: 10.2337/db16-0924.

8 É. Szentirmai, N. S. Millican, A. R. Massie, et al. "Butyrate, a Metabolite of Intestinal Bacteria, Enhances Sleep." *Sci Rep* 9(1) (2019): 7035. doi: 10.1038/s41598-019-43502-1.

9 S. M. Matt, J. M. Allen, M. A. Lawson, L. J. Mailing, J. A. Woods, and R. W. Johnson. "Butyrate and Dietary Soluble Fiber Improve Neuroinflammation Associated with Aging in Mice." *Front Immunol* 9 (2018): 1832. doi: 10.3389/fimmu.2018.01832.

10 R. Mastrocola, F. Restivo, I. Vercellinatto, O. Danni, E. Brignardello, M. Aragno, and G. Boccuzzi. "Oxidative and Nitrosative Stress in Brain Mitochondria of Diabetic Rats." *J Endocrinol* 187(1) (2005): 37–44. doi: 10.1677/joe.1.06269.

11 A. Czajka and A. N. Malik. "Hyperglycemia Induced Damage to Mitochondrial Respiration in Renal Mesangial and Tubular Cells: Implications for Diabetic Nephropathy." *Redox Biol* 10 (2016): 100–107. doi: 10.1016/j.redox.2016.09.007.

12 A. J. Sommerfield, I. J. Deary, and B. M. Frier. "Acute Hyperglycemia Alters Mood State and Impairs Cognitive Performance in People with Type 2 Diabetes." *Diabetes Care* 27(10) (2004): 2335–2340. doi: 10.2337/diacare.27.10.2335.

13 M. Kirvalidze, A. Hodkinson, D. Storman, T. J. Fairchild, M. M. Bała, G. Beridze, A. Zuriaga, N. I. Brudasca, and S. Brini. "The Role of Glucose in Cognition, Risk of Dementia, and Related Biomarkers in Individuals Without Type 2 Diabetes Mellitus or the Metabolic Syndrome: A Systematic Review of Observational Studies." *Neurosci Biobehav* 135 (Rev. April 2022): 104551. doi: 10.1016/j.neubiorev.2022.104551.

14 C. Toda, J. D. Kim, D. Impellizzeri, S. Cuzzocrea, Z. W. Liu, S. Diano. "UCP2 Regulates Mitochondrial Fission and Ventromedial Nucleus Control of Glucose Responsiveness." *Cell* 164(5) (2016): 872–883. doi: 10.1016/j.cell.2016.02.010.

15 A. Fagiolini, D. J. Kupfer, P. R. Houck, D. M. Novick, and E. Frank. "Obesity as a Correlate of Outcome in Patients with Bipolar I Disorder." *Am J Psychiatry* 160(1) (2003): 112–117. doi: 10.1176/appi.ajp.160.1.112.

16 Noppamas Pipatpiboon, Wasana Pratchayasakul, Nipon Chattipakorn, and Siriporn C. Chattipakorn. "PPARγ Agonist Improves Neuronal Insulin Receptor Function in Hippocampus and Brain Mitochondria Function in Rats with Insulin Resistance Induced by Long Term High-Fat Diets." *Endocrinology* 153(1) (2012): 329–338. doi: 10.1210/en.2011-1502.

17 H. Y. Liu, E. Yehuda-Shnaidman, T. Hong, et al. "Prolonged Exposure to Insulin Suppresses Mitochondrial Production in Primary Hepatocytes." *J Biol Chem* 284(21) (2009): 14087–14095. doi: 10.1074/jbc.M807992200.

18 K. Wardelmann, S. Blümel, M. Rath, E. Alfine, C. Chudoba, M. Schell, W. Cai, R. Hauffe, K. Warnke, T. Flore, K. Ritter, J. Weiß, C. R. Kahn, and A. Kleinridders. "Insulin Action in the Brain Regulates Mitochondrial Stress Responses and Reduces Diet-Induced Weight Gain." *Mol Metab* 21(2019): 68–81. doi: 10.1016/j.molmet.2019.01.001.

19 J. D. Kim, N. A. Yoon, S. Jin, and S. Diano. "Microglial UCP2 Mediates Inflammation and Obesity Induced by High-Fat Feeding." *Cell Metab* 30(5) (2019): 952–962. e5. doi: 10.1016/j.cmet.2019.08.010.

20 M. O. Dietrich, Z. W. Liu, and T. L. Horvath. "Mitochondrial Dynamics Controlled by Mitofusins Regulate Agrp Neuronal Activity and Diet-Induced Obesity." *Cell* 155(1) (2013): 188-199. doi: 10.1016/j.cell.2013.09.004; M. Schneeberger, M. O. Dietrich, D. Sebastián, et al. "Mitofusin 2 in POMC Neurons Connects ER Stress with Leptin Resistance and Energy Imbalance." *Cell* 155(1) (2013): 172–187. doi: 10.1016/j.cell.2013.09.003.

21 A. S. Rambold, B. Kostelecky, N. Elia, and J. Lippincott-Schwartz. "Tubular Network Formation Protects Mitochondria from Autophagosomal Degradation During Nutrient Starvation." *Proc Natl Acad Sci USA* 108(25) (2011): 10190–10195. doi: 10.1073/pnas.1107402108.

22 A. Keys, J. Brozek, A. Henshel, O. Mickelson, and H. L. Taylor. *The Biology of Human Starvation*, vols. 1–2 (Minneapolis: University of Minnesota Press, 1950).

23 C. Lindfors, I. A. Nilsson, P. M. Garcia-Roves, A. R. Zuberi, M. Karimi, L. R. Donahue, D. C. Roopenian, J. Mulder, M. Uhlén, T. J. Ekström, M. T. Davisson, T. G. Hökfelt, M. Schalling, and J. E. Johansen. "Hypothalamic Mitochondrial Dysfunction Associated with Anorexia in the Anx/Anx Mouse." *Proc Natl Acad Sci USA* 108(44) (2011): 18108–18113. doi: 10.1073/pnas.1114863108.

24 V. M. Victor, S. Rovira-Llopis, V. Saiz-Alarcon, et al. "Altered Mitochondrial Function and Oxidative Stress in Leukocytes of Anorexia Nervosa Patients." *PLoS One* 9(9) (2014): e 106463. doi: 10.1371/journal.pone.0106463.

25 P. Turnbaugh, R. Ley, M. Mahowald, et al. "An Obesity-Associated Gut Microbiome with Increased Capacity for Energy Harvest." *Nature* 444(7122) (2006): 1027–1031. doi: 10.1038/nature05414.

26 D. N. Jackson and A. L. Theiss. "Gut Bacteria Signaling to Mitochondria in Intestinal Inflammation and Cancer." *Gut Microbes* 11(3) (2020): 285–304. doi: 10.1080/19490976.2019.1592421.

27 C. M. Palmer. "Diets and Disorders: Can Foods or Fasting Be Considered Psychopharmacologic Therapies?" *J Clin Psychiatry* 81(1) (2019): 19ac12727. doi: 10.4088/JCP.19ac12727. PMID: 31294934.

28 C. T. Hoepner, R. S. McIntyre, and G. I. Papakostas. "Impact of Supplementation and Nutritional Interventions on Pathogenic Processes of Mood Disorders: A Review of the Evidence." *Nutrients* 13(3) (2021): 767. doi: 10.3390/nu13030767; National Institutes of Health, Office of Dietary Supplements. June 3, 2020. "Dietary Supplements for Primary Mitochondrial Disorders." NIH, https://ods.od.nih.gov/factsheets/PrimaryMitochondrialDisorders-HealthProfessional/. Retrieved 7/24/21.

29 M. Berk, A. Turner, G. S. Malhi, et al. "A Randomised Controlled Trial of a Mitochondrial Therapeutic Target for Bipolar Depression: Mitochondrial Agents, N-acetylcysteine, and Placebo." *BMC Med* 17(1) (2019): 18. [Published correction appears in *BMC Med* 17(1) (2019): 35.] doi:10.1186/s12916-019-1257-1.

30 F. N. Jacka, A. O'Neil, R. Opie, et al. "A Randomised Controlled Trial of Dietary Improvement for Adults with Major Depression (the 'SMILES' Trial)." *BMC Med* 15(1) (2017): 23. doi: 10.1186/s12916-017-0791-y.

31 K. A. Amick, G. Mahapatra, J. Bergstrom, Z. Gao, S. Craft, T. C. Register, C. A. Shively, and A. J. A. Molina. "Brain Region-Specific Disruption of Mitochondrial Bioenergetics in Cynomolgus Macaques Fed a Western Versus a Mediterranean Diet." *Am J Physiol Endocrinol Metab* 321(5) (2021): E652–E664. doi: 10.1152/ajpendo.00165.2021.

32 Y. Liu, A. Cheng, Y. J. Li, Y. Yang, Y. Kishimoto, S. Zhang, Y. Wang, R. Wan, S. M. Raefsky, D. Lu, T. Saito, T. Saido, J. Zhu, L. J. Wu, and M. P. Mattson. "SIRT3 Mediates Hippocampal Synaptic Adaptations to Intermittent Fasting and Ameliorates Deficits in APP Mutant Mice." *Nat Commun* 10(1) (2019): 1886. doi: 10.1038/s41467-019-09897-1.

33 M. Mattson, K. Moehl, N. Ghena, et al. "Intermittent Metabolic Switching, Neuroplasticity and Brain Health." *Nat Rev Neurosci* 19(2) (2018): 81–94. doi: 10.1038/nrn.2017.156.

34 K. J. Martin-McGill, R. Bresnahan, R. G. Levy, and P. N. Cooper. "Ketogenic Diets for Drug-Resistant Epilepsy." *Cochrane Database Syst Rev* 6(6) (2020): CD001903. doi: 10.1002/14651858.CD001903.pub5.

35 K. J. Bough, J. Wetherington, B. Hassel, J. F. Pare, J. W. Gawryluk, J. G. Greene, R. Shaw, Y. Smith, J. D. Geiger, and R. J. Dingledine. "Mitochondrial Biogenesis in the Anticonvulsant Mechanism of the Ketogenic Diet." *Ann Neurol* 60(2) (2006): 223–235. doi: 10.1002/ana.20899; J. M. Rho. "How Does the Ketogenic Diet Induce Anti-Seizure Effects?" *Neurosci Lett* 637 (2017): 4–10. doi: 10.1016/j.neulet.2015.07.034.

36 C. M. Palmer, J. Gilbert-Jaramillo, and E. C. Westman. "The Ketogenic Diet and Remission of Psychotic Symptoms in Schizophrenia: Two Case Studies." *Schizophr Res* 208 (2019): 439–440. doi: 10.1016/j.schres.2019.03.019. Epub April 6, 2019. PMID: 30962118.

37 M. C. L. Phillips, L. M. Deprez, G. M. N. Mortimer, et al. "Randomized Crossover Trial of a Modified Ketogenic Diet in Alzheimer's Disease." *Alzheimer's Res Ther* 13(1) (2021): 51. doi: 10.1186/s13195-021-00783-x.

Chapter 16

1 H. K. Seitz, R. Bataller, H. Cortez-Pinto, B. Gao, A. Gual, C. Lackner, P. Mathurin, S. Mueller, G. Szabo, and H. Tsukamoto. "Alcoholic Liver Disease." *Nat Rev Dis Primers* 4(1) (2018): 16. doi: 10.1038/s41572-018-0014-7. Erratum in: *Nat Rev Dis Primers* 4(1) (2018): 18. PMID: 30115921.

2 C. Tapia-Rojas, A. K. Torres, and R. A. Quintanilla. "Adolescence Binge Alcohol Consumption Induces Hippocampal Mitochondrial Impairment That Persists During the Adulthood." *Neuroscience* 406 (2019): 356–368. doi: 10.1016/j.neuroscience.2019.03.018.

3 Nora D. Volkow, Sung Won Kim, Gene-Jack Wang, David Alexoff, Jean Logan, Lisa Muench, Colleen Shea, et al. "Acute Alcohol Intoxication Decreases Glucose Metabolism but Increases Acetate Uptake in the Human Brain." *NeuroImage* 64 (2013): 277–283. doi: 10.1016/j.neuroimage.2012.08.057.

4 N. D. Volkow, G. J. Wang, E. Shokri Kojori, J. S. Fowler, H. Benveniste, and D. Tomasi. "Alcohol Decreases Baseline Brain Glucose Metabolism More in Heavy Drinkers Than Controls but Has No Effect on Stimulation-Induced Metabolic Increases." *J Neurosci* 35(7) (2015): 3248–3255. doi:10.1523/JNEUROSCI.4877-14.2015.

5 C. E. Wiers, L. F. Vendruscolo, J. W. van der Veen, et al. "Ketogenic Diet Reduces Alcohol Withdrawal Symptoms in Humans and Alcohol Intake in Rodents." *Sci Adv* 7(15) (2021): eabf6780. doi: 10.1126/sciadv.abf6780.

6 N. D. Volkow, J. M. Swanson, A. E. Evins, L. E. DeLisi, M. H. Meier, R. Gonzalez, M. A. Bloomfield, H. V. Curran, and R. Baler. "Effects of Cannabis Use on Human Behavior, Including Cognition, Motivation, and Psychosis: A Review." *JAMA Psychiatry* 73(3) (2016): 292–297. doi: 10.1001/jamapsychiatry.2015.3278.

7 T. Harkany and T. L. Horvath. "(S)Pot on Mitochondria: Cannabinoids Disrupt Cellular Respiration to Limit Neuronal Activity." *Cell Metab* 25(1) (2017): 8–10. doi: 10.1016/j.cmet.2016.12.020.

8 M. D. Albaugh, J. Ottino-Gonzalez, A. Sidwell, et al. "Association of Cannabis Use During Adolescence with Neurodevelopment." *JAMA Psychiatry* 78(9) (2021): 1031–1040. doi: 10.1001/jamapsychiatry.2021.1258.

9 D. Jimenez-Blasco, A. Busquets-Garcia, et al. "Glucose Metabolism Links Astroglial Mitochondria to Cannabinoid Effects." *Nature* 583(7817) (2020): 603–608. doi: 10.1038/s41586-020-2470-y.

10 E. Hebert-Chatelain, T. Desprez, R. Serrat, et al. "A Cannabinoid Link Between Mitochondria and Memory." *Nature* 539(7630) (November 24, 2016): 555–559. doi: 10.1038/nature20127.

Chapter 17

1 S. R. Chekroud, R. Gueorguieva, A. B. Zheutlin, M. Paulus, H. M. Krumholz, J. H. Krystal, and A. M. Chekroud. "Association Between Physical Exercise and Mental Health in 1.2 Million Individuals in the USA Between 2011 and 2015: A Cross-Sectional Study. *Lancet Psychiatry* 5(9) (2018): 739–746. doi: 10.1016/S2215-0366(18)30227-X.

2 G. A. Greendale, W. Han, M. Huang, et al. "Longitudinal Assessment of Physical Activity and Cognitive Outcomes Among Women at Midlife." *JAMA Netw Open* 4(3) (2021): e213227. doi: 10.1001/jamanetworkopen.2021.3227.

3 J. Krogh, C. Hjorthøj, H. Speyer, C. Gluud, and M. Nordentoft. "Exercise for Patients with Major Depression: A Systematic Review with Meta-Analysis and Trial Sequential Analysis." *BMJ Open* 7(9) (2017): e014820. doi: 10.1136/bmjopen-2016-014820.

4 World Health Organization. *Motion for Your Mind: Physical Activity for Mental Health Promotion, Protection, and Care.* Copenhagen: WHO Regional Office for Europe, 2019. https://www.euro.who.int/en/health-topics/disease-prevention/physical-activity/publications/2019/motion-for-your-mind-physical-activity-for-mental-health-promotion,-protection-and-care-2019.

5 K. Contrepois, S. Wu, K. J. Moneghetti, D. Hornburg, et al. "Molecular Choreography of Acute Exercise." *Cell* 181(5) (2020): 1112–1130.e16. doi: 10.1016/j.cell.2020.04.043.

6 A. R. Konopka, J. L. Laurin, H. M. Schoenberg, J. J. Reid, W. M. Castor, C. A. Wolff, R. V. Musci, O. D. Safairad, M. A. Linden, L. M. Biela, S. M. Bailey, K. L. Hamilton, and B. F. Miller. "Metformin Inhibits Mitochondrial Adaptations to Aerobic Exercise Training in Older Adults." *Aging Cell* 18(1) (2019): e12880. doi: 10.1111/acel.12880.

7 Kathrin Steib, Iris Schäffner, Ravi Jagasia, Birgit Ebert, and D. Chichung Lie. "Mitochondria Modify Exercise-Induced Development of Stem Cell-Derived Neurons in the Adult Brain." *J Neurosci* 34(19) (2014): 6624. doi: 10.1523/JNEUROSCI.4972-13.2014.

Chapter 18

1 H. T. Chugani, M. E. Behen, O. Muzik, Ç. Juhász, F. Nagy, and D. C. Chugani. "Local Brain Functional Activity Following Early Deprivation: A Study of Postinstitutionalized Romanian Orphans." *Neuroimage* 14(6) (2001): 1290–1301. doi: 10.1006/nimg.2001.0917.

2 M. Picard, A. A. Prather, E. Puterman, A. Cuillerier, M. Coccia, K. Aschbacher, Y. Burelle, and E. S. Epel. "A Mitochondrial Health Index Sensitive to Mood and Caregiving Stress." *Biol Psychiatry* 84(1) (2018): 9–17. doi: 10.1016/j.biopsych.2018.01.012.

3 Frankl, V. E. *Man's Search for Meaning: An Introduction to Logotherapy* (New York: Simon & Schuster, 1984).

4 A. Alimujiang, A. Wiensch, J. Boss, et al. "Association Between Life Purpose and Mortality Among US Adults Older Than 50 Years." *JAMA Netw Open* 2(5) (2019): e194270. doi: 10.1001/jamanetworkopen.2019.4270.

5 R. Cohen, C. Bavishi, and A. Rozanski. "Purpose in Life and Its Relationship to All-Cause Mortality and Cardiovascular Events: A Meta-Analysis." *Psychosom Med* 78(2) (2016): 122–133. doi: 10.1097/PSY.0000000000000274.

6 L. Miller, R. Bansal, P. Wickramaratne, et al. "Neuroanatomical Correlates of Religiosity and Spirituality: A Study in Adults at High and Low Familial Risk for Depression." *JAMA Psychiatry* 71(2) (2014): 128–135. doi: 10.1001/jamapsychiatry.2013.3067.

7 T. J. VanderWeele, S. Li, A. C. Tsai, and I. Kawachi. "Association Between Religious Service Attendance and Lower Suicide Rates Among US Women." *JAMA Psychiatry* 73(8) (2016): 845–851. doi: 10.1001/jamapsychiatry.2016.1243.

8 H. G. Koenig. "Religion, Spirituality, and Health: The Research and Clinical Implications." *ISRN Psychiatry* 2012 (2012): 278730. doi: 10.5402/2012/278730

9 C. Timmermann, H. Kettner, C. Letheby, et al. "Psychedelics Alter Metaphysical Beliefs." *Sci Rep* 11(1) (2021): 22166. doi: 10.1038/s41598-021-01209-2.

10 J. A. Dusek, H. H. Otu, A. L. Wohlhueter, M. Bhasin, L. F. Zerbini, M. G. Joseph, H. Benson, and T. A. Libermann. "Genomic Counter-Stress Changes Induced by the Relaxation Response." *PLoS One* 3(7) (2008): e2576. doi: 10.1371/journal.pone.0002576.

11 M. K. Bhasin, J. A. Dusek, B. H. Chang, M. G. Joseph, J. W. Denninger, G. L. Fricchione, H. Benson, and T. A. Libermann. "Relaxation Response Induces Temporal Transcriptome Changes in Energy Metabolism, Insulin Secretion and Inflammatory Pathways." *PLoS One* 8(5) (2013): e62817. doi: 10.1371/journal.pone.0062817.

Chapter 19

1 M. Búrigo, C. A. Roza, C. Bassani, D. A. Fagundes, G. T. Rezin, G. Feier, F. Dal-Pizzol, J. Quevedo, and E. L. Streck. "Effect of Electroconvulsive Shock on Mitochondrial Respiratory Chain in Rat Brain." *Neurochem Res* 31(11) (2006): 1375–1379. doi: 10.1007/s11064-006-9185-9.

2 F. Chen, J. Danladi, G. Wegener, T. M. Madsen, and J. R. Nyengaard. "Sustained Ultrastructural Changes in Rat Hippocampal Formation After Repeated Electroconvulsive Seizures." *Int J Neuropsychopharmacol* 23(7) (2020): 446–458. doi: 10.1093/ijnp/pyaa021.

3 F. J. Medina and I. Túnez. "Mechanisms and Pathways Underlying the Therapeutic Effect of Transcranial Magnetic Stimulation." *Rev Neurosci* 24(5) (2013): 507–525. doi: 10.1515/revneuro-2013-0024.

4 H. L. Feng, L. Yan, and L. Y. Cui. "Effects of Repetitive Transcranial Magnetic Stimulation on Adenosine Triphosphate Content and Microtubule Associated Protein-2 Expression After Cerebral Ischemia-Reperfusion Injury in Rat Brain." *Chin Med J* (Engl) 121(14) (2008): 1307–1312. PMID: 18713553.

5 X. Zong, Y. Dong, Y. Li, L. Yang, Y. Li, B. Yang, L. Tucker, N. Zhao, D. W. Brann, X. Yan, S. Hu, and Q. Zhang. "Beneficial Effects of Theta-Burst Transcranial Magnetic Stimulation on Stroke Injury via Improving

Neuronal Microenvironment and Mitochondrial Integrity." *Transl Stroke Res* 11(3) (2020): 450–467. doi: 10.1007/s12975-019-00731-w.

6 C. L. Cimpianu, W. Strube, P. Falkai, U. Palm, and A. Hasan. "Vagus Nerve Stimulation in Psychiatry: A Systematic Review of the Available Evidence." *J Neural Transm* (Vienna) 124(1) (2017): 145–158. doi: 10.1007/s00702-016-1642-2.

Index

D

N

T

U

V

W

Y

About the Author

Christopher M. Palmer, MD, is a Harvard psychiatrist and researcher working at the interface of metabolism and mental health. He is the director of the Department of Postgraduate and Continuing Education at McLean Hospital and an assistant professor of psychiatry at Harvard Medical School. For more than two decades, he has held leadership roles in psychiatric education at Harvard, McLean Hospital, and nationally. He spent more than fifteen years conducting neuroscience research in the areas of substance use and sleep disorders. On top of these academic pursuits, he has continued to practice psychiatry, working with people who have treatment-resistant mental disorders using a variety of standard treatments. He has been pioneering the use of the medical ketogenic diet in the treatment of psychiatric disorders—conducting research in this area, treating patients, publishing academic articles, and speaking globally on this topic. Most recently, he has developed the first comprehensive theory of what causes mental illness, integrating biological, psychological, and social research into one unifying theory—the brain energy theory of mental illness.